KW-421-752

APPLIED
MECHANICS

APPLIED MECHANICS

BY

P. D. COLLINS

M.Sc., A.M.I.Mech.E., A.M.I.Prod.E.

*Senior Lecturer in Mechanical Engineering
at the S.-E. Essex Technical College*

LONGMANS

LONGMANS, GREEN AND CO LTD
6 & 7 CLIFFORD STREET, LONDON W I

THIBAULT HOUSE, THIBAULT SQUARE, CAPE TOWN
605–611 LONSDALE STREET, MELBOURNE C I
443 LOCKHART ROAD, HONG KONG
ACCRA, AUCKLAND, IBADAN
KINGSTON (JAMAICA), KUALA LUMPUR
LAHORE, NAIROBI, SALISBURY (RHODESIA)

LONGMANS, GREEN AND CO INC
119 WEST 40TH STREET, NEW YORK 18

LONGMANS, GREEN AND CO
20 CRANFIELD ROAD, TORONTO 16

ORIENT LONGMANS PRIVATE LTD
CALCUTTA, BOMBAY, MADRAS
DELHI, HYDERABAD, DACCA

© P. D. Collins 1960

First published 1960

PRINTED IN GREAT BRITAIN
BY R. & R. CLARK, LTD., EDINBURGH

PREFACE

THE scope of this book has been limited to the main topics which must normally be covered by students taking the final year of a part-time course of study in preparation for the Ordinary National Certificate examination in Applied Mechanics.

A considerable number of worked examples have been included in the text and it is hoped that these will be particularly helpful to students working independently for the Part I examination in Applied Mechanics of the Professional Engineering Institutions.

Answers are provided to the exercises which are set at the end of each chapter, and where additional guidance is thought to be necessary, part of the solution is also given.

For the student who will be proceeding to the Higher National Certificate course, an attempt has been made to point out the way in which some of the more important sections of the work will be developed at a later stage.

I wish to thank the Institution of Mechanical Engineers, the Institution of Production Engineers, the University of London, the Northern Counties Technical Examination Council, the Union of Educational Institutions and the Union of Lancashire and Cheshire Institutes for permission to use questions from their examination papers.

I also wish to thank Mr. R. S. Paradise, B.Sc.(Eng.), M.I.Mech.E., for his very helpful advice and criticism, and Mr. R. E. Hanley, B.Sc.(Eng.), A.M.I.Mech.E., for checking the solutions.

ILFORD,   P. D. COLLINS.
  ESSEX.

v

CONTENTS

vii

Chapter I

STATICS

WHEN a body is in equilibrium under the action of co-planar forces which are either *concurrent* (pass through one point) or are parallel, the force and funicular polygons *both* close.

In the general case of co-planar *non-concurrent* forces the resultant force on the body may be zero, i.e. the force polygon closes, but the body need not be in equilibrium. In such cases the funicular polygon does not close and equilibrium can only be achieved by the application of a couple.

1. Force Polygon.—When a number of co-planar forces act on a body the 'polygon of forces' may be used to determine the magnitude, direction and sense of direction of their resultant.

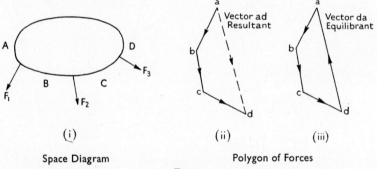

(i)

(ii)   (iii)

Space Diagram   Polygon of Forces

FIG. 1

Consider the three forces F_1, F_2 and F_3 acting on the body in Fig. 1 (i). This figure, which indicates the physical arrangement of the forces, is called the 'space diagram'. The force polygon at (ii) is drawn according to Bow's notation, which consists of specifying the forces in the space diagram by labelling the spaces between them with capital letters and using corresponding small letters to represent the force vectors, e.g. Force F_1 is denoted by

AB on the space diagram and its vector by '*ab*' on the force polygon.

From the force polygon it is seen that the resultant of the vectors *ab*, *bc* and *cd* is given by vector *ad* in magnitude and direction.

If, now, a force equal and opposite to the resultant acts on the body, the force polygon closes as in (iii), i.e. the arrowheads follow round the diagram and there is no resultant force on the body. The closing vector *da*, therefore, gives the equilibrant force in magnitude and direction.

2. Funicular or Link Polygon.
—The force polygon gives the magnitude and direction of the resultant (and equilibrant) of a system of forces but does not indicate its relative position on the space diagram.

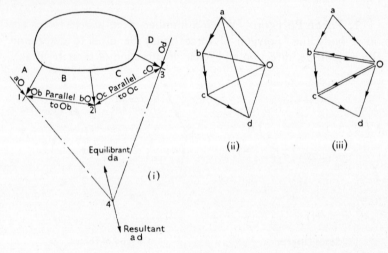

Fig. 2

To determine this position the funicular or link polygon is used as outlined below.

Choosing any pole point O at a convenient distance from the force polygon, draw the lines O*a*, O*b*, etc., as in Fig. 2 (ii).

In the triangle O*ab* the vector *ab* may be considered as the resultant of forces represented by *a*O and O*b*—or, in other words—the effect of the force *ab* may be replaced by components represented by aO and Ob. Similarly for the vectors bc and cd, as shown at (iii), where the component triangles are illustrated separately. This diagram also shows that the components bO and Ob cancel themselves, which is also the case with cO and Oc, thereby leaving only the components aO and Od to represent the three actual forces on the body. Therefore, these three forces may be replaced by the two components aO and Od which have ad as their resultant vector.

Now at any point (1) on the line of action of the force AB draw component lines parallel to aO and Ob. Similarly for the force BC, noting that the lines Ob and bO coincide in space B, and so on round the diagram.

Produce the lines of action of components aO and Od to intersect at point (4). The resultant of these two components will pass through this point and will be parallel to ad on the force polygon. The closed figure 1-2-3-4 forms the funicular polygon. It is important to note that *closure* of the funicular polygon indicates that equilibrium can be obtained by the addition of the single force da—the equilibrant. For a body to be in equilibrium not only does the force polygon close but the funicular polygon closes as well. Closure of the force polygon alone only shows that there is no resultant force acting on a body—it does not indicate that the body is in equilibrium since there may be a resulting moment or couple acting on the body (see example 3).

3. Graphical Determination of Beam Reactions.—It is required to determine the reactions at the ends of a beam carrying parallel loads by means of the funicular polygon—Fig. 3. The weight of the beam is neglected.

A force diagram consisting of a straight line $abcd$ is constructed at (ii) and a suitable pole O is chosen. The funicular polygon may be commenced at any convenient point (i) on the line of the left-hand reaction ; 1-2 is drawn in space A parallel to aO in the force polygon ; 2-3 in space B parallel to bO, and so on, until point 5 is located on the line of action of the right-hand support. Points 1 and 5 are joined in the space E, and eO drawn parallel to it on the force polygon. From the previous discussion on component triangles, the force represented by aO is equivalent to the vertical force ae and a force eO along the closing line 1-5 of the funicular polygon. Similarly at the right-hand end, Od is equivalent to *ed* and O*e*—the components *e*O and O*e* cancelling in this diagram. The three forces AB, BC and CD can, therefore, be replaced by forces *ae* and *ed*, and hence the forces required to

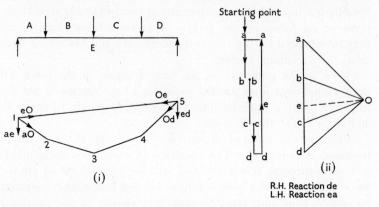

Fig. 3

R.H. Reaction de
L.H. Reaction ea

balance them must be equal and opposite—i.e. the reactions are *de* at the right-hand end and *ea* at the left.

Example 1

Determine the reactions for the beam given in Fig. 4 by means of the funicular polygon.

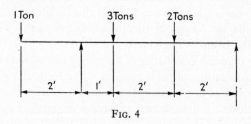

Fig. 4

Solution

The space diagram and force polygon *abcd* are drawn to scale in Fig. 5. The funicular polygon is commenced at point 1 and the line 1-2 is drawn parallel to O*b right across* space B ; 2-3 in space C parallel to O*c*; and 3-4 parallel to O*d* in space D, and terminating at 4 in the right-hand reaction line.

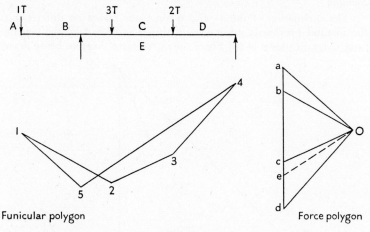

Funicular polygon   Force polygon

FIG. 5

Line 1-5 terminates on the left-hand reaction line and is parallel to Oa.

The closing line 4-5 is transferred to the force polygon as Oe and gives the reactions as de and ea.

Left-hand Reaction $ea = 4.6$ tons,

Right-hand Reaction $de = 1.4$ tons.

Example 2

The horizontal beam XY shown in Fig. 6 is hinged at X and simply supported at Y. Using the funicular polygon determine :

(*a*) The magnitude and direction of the reaction at the hinge X.

(*b*) The magnitude, direction and point of application of the resultant load on the beam.

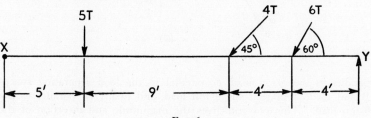

FIG. 6

Solution

The inclination of the 4- and 6-ton loads makes no difference to the method previously adopted and this is indicated by the separate force diagram shown at (ii) Fig. 7, *ab* in the force diagram being drawn

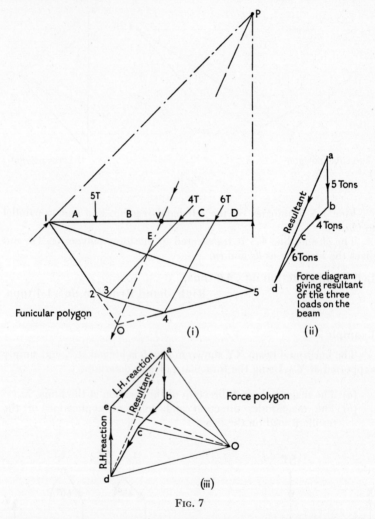

Fig. 7

parallel to the 5-ton load, *bc* parallel to the load separating spaces B and C, to represent the 4-ton load, and finally *cd* parallel to the 6-ton load. This polygon gives the magnitude and direction of the resultant of the three loads as vector '*ad*'.

All that is known about the hinge reaction at A is that it must pass through the point (1) at the end of the beam. For this reason, the funicular polygon is commenced at the hinge—the remaining construction following as previously explained.

The closing line 1-5 over the space E is transferred to the force polygon through O to meet a vertical through *d*, parallel with the right-hand reaction at *e*. Then the resultant '*ad*', together with the two reactions, forms the triangle of forces *ade*, in which the direction and magnitude of the hinge reaction is given by '*ea*'. This is transferred to the space diagram through the point (1).

To determine the direction and point of application of the resultant load on the beam, lines 1-2 and 5-4 of the funicular polygon are produced to intersect at Q. The resultant load will then pass through point Q—its direction being parallel to the vector *ad* in the force diagram. Hence the point on the beam through which the resultant load will act is given by V.

Alternatively :

Since the beam is in equilibrium, it is known by the triangle of forces that the resultant force '*ad*' must pass through P—the intersection of the line of action of the two end reactions.

Reaction at A :	Magnitude	**8·3 tons**
	Direction	**43·3° to Horizontal**
Resultant :	Magnitude	**14·3 tons**
	Direction	**65° to Horizontal**
	Point of Application	**12·4 ft. from L.H.E.**

Example 3

By using the funicular polygon determine the magnitude of the reaction supplied by the wall for the cantilever beam shown in Fig. 8.

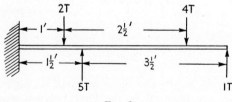

Fig. 8

Solution

The net vertical force is zero—by inspection.

Hence the force polygon is a closed figure and the resultant *force* on the beam is zero.

The force polygon *abcde* is drawn to scale as shown in Fig. 9 (ii), and the funicular polygon commenced at any point (i) on the line of action of the force AB.

Line 1-2 is parallel to O*b*; 2-3 parallel to O*c*, meeting the line of action of the 1-ton force produced in 3.

Line 3-4 is drawn right across space D parallel to O*d*.

Lines parallel to O*a* and O*e* in the force polygon are drawn from

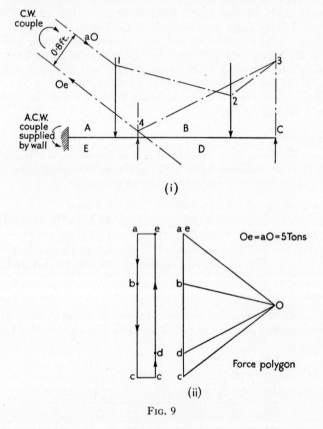

(i)

Fig. 9

1 and 4, respectively, to give the line of action of the forces *a*O and O*e*. These two lines are identical in the force polygon, and hence are parallel in the funicular diagram.

Therefore, the funicular polygon does not close.

Clearly these two equal and opposite forces form a couple. Hence the resultant action of the forces on the beam is a clockwise couple of magnitude 5 tons × 0·8 feet = **4·0 ton ft.**

For equilibrium the wall fixing must supply an equal and opposite reaction couple as illustrated in the diagram.

This result can easily be checked by the principle of moments.

4. Simple Framed Structures.

4. Simple Framed Structures.—The design of any structure, such as a space frame or a plane roof truss which is made up of members joined at their ends, involves the determination of the forces in each member under a given system of loading. These forces can be determined for 'simple structures' providing two assumptions are made which will first be discussed in some detail.

'All the connections between individual members in a simple structure will be considered as pin-jointed.'

When a member is pinned at the end, it is free to rotate, and, therefore, there can be no moment at the joint. Hence the significance of the term pin-joint—in a pin-jointed frame there is no bending in the members.

Fig. 10(a) illustrates the usual type of riveted connection

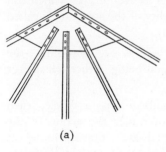

(a)

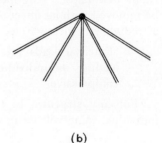

(b)

Fig. 10

between members which, for ease of calculation, is replaced by the pin-joint shown at (b).

The term 'simple structure' applies to those frames which are made up of individual triangular elements as shown in Fig. 11.

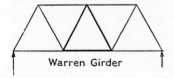

Warren Girder

Pratt or N Girder

Fig. 11

A.M.—B

Since each individual triangle—pin-jointed at the corners—is completely rigid, any frame built up in this way is also rigid and will not collapse.

Consider now the frame in Fig. 12. There are just sufficient members to give triangulation, and, therefore, the frame is rigid.

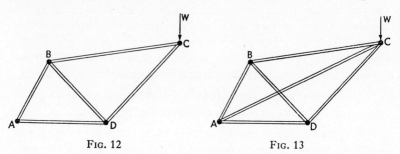

FIG. 12   FIG. 13

Removal of the member BD will destroy the equilibrium and the frame will then collapse under load. This is not the case in Fig. 13 where the removal of either AC or BD will not cause the frame to collapse. Any member, such as AC or BD, which can be removed without affecting the rigidity of the structure is called a 'redundant' member (see example 4).

> 'External loads are only applied at the pin-joints—the loading on individual members being either tensile or compressive.'

For any structure to be in equilibrium as a whole, each individual joint will be in equilibrium under the action of the external and internal forces at that joint. In the simple frame of Fig. 14 the members AB and AC are clearly in compression (Struts). For equilibrium at pin A, these members will push outwards on the pin, and the arrowheads *on* them are those of the *internal* forces. Member BC is in tension (Tie) and the internal forces pull inwards on the pins at B and C.

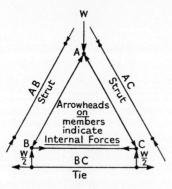

FIG. 14

There are three methods

available for the determination of forces in simple structures :

(i) Method of Joints,
(ii) Force Diagram,
(iii) Method of Sections.

Methods (i) and (iii) are useful where the forces are required in particular members of a frame only, whereas the Force Diagram method enables the force in every member of a frame to be obtained from a single diagram.

Each of these methods will now be described with reference to the framework shown in Fig. 15.

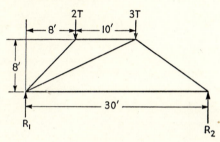

Fig. 15

In this problem a convenient first step for all three methods is to determine the reactions R_1 and R_2.

This may be done either by taking moments about one support or by means of the funicular polygon.

Reactions

By Moments about R_1 $CW = (2 \times 8) + (3 \times 18) = 70$ ton ft.,
$$ACW = 30 R_2,$$
$$\therefore R_2 = 2\tfrac{1}{3} \text{ tons}, \qquad R_1 = 2\tfrac{2}{3} \text{ tons}.$$

(i) *Method of Joints*

This consists of determining the forces which give equilibrium at each joint of the frame in turn.

They may be obtained either graphically by the polygon of forces or analytically by satisfying the equations for static equilibrium, viz.

Σ Vertical Forces = zero. Σ Horizontal Forces = zero.

Polygon of Forces. (See Fig. 16.)

The space diagram (*a*) is labelled according to Bow's notation. It is necessary to begin at one of the joints where there are not

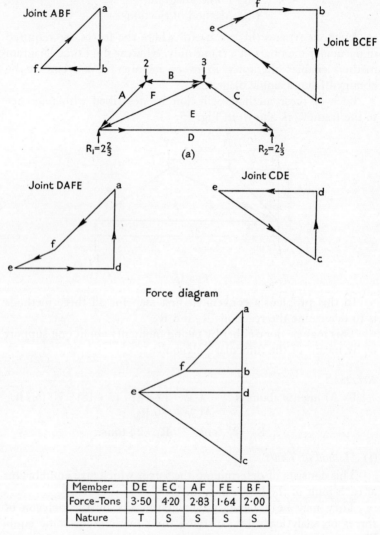

Member	DE	EC	AF	FE	BF
Force-Tons	3·50	4·20	2·83	1·64	2·00
Nature	T	S	S	S	S

Fig. 16

more than two unknown forces, i.e. the internal forces in the members at that joint.

(a) Joint CDE

This gives the triangle *cde* which is obtained by following round the joint in the clockwise direction C-D-E. The arrows which follow round this triangle indicate that ED is a tie—i.e. it pulls inwards on the joint.

It follows that ED also pulls inwards at the left-hand reaction.

ED = 3·5-tons Tie,   EC = 4·2-tons Strut.

(b) Joint DAFE

This polygon can be constructed since there are now two known forces at this joint. The clockwise direction round the joint being D-A-F-E.

AF = 2·83-tons Strut,   FE = 1·64-tons Strut.

(c) Joint ABF

The triangle of forces at this joint—which could have been taken as the starting-point—gives

BF = 2·0-tons Strut.

For this simple example no separate polygon is actually required for the joint BCEF.

Analytically :
(a) Joint CDE

Member EC must supply a vertically downward component to balance R_2. This requires that the force in ED shall balance the horizontal component of EC.

Resolving forces :
Vertically   EC sin $\alpha = 2\frac{1}{3}$,
                EC = 4·2-tons Strut,
Horizontally EC cos α = ED,
                ED = 3·5-tons Tie.

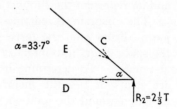

(b) Joint BCEF

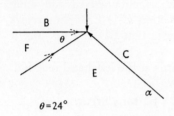

$\theta = 24°$

Resolving forces :
Vertically EC sin α + FE sin θ = 3,

$$2\tfrac{1}{3} + \text{FE sin } 24° = 3,$$

FE = 1·64-tons Strut.

Horizontally EC cos α = FE cos θ + BF,
$$3·5 = 1·64 \cos 24° + \text{BF},$$
BF = 2·0-tons Strut.

(c) Joint ABF

The only unknown member at this joint is now AF.

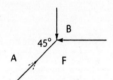

Resolving forces :
Horizontally AF cos 45° = BF,

$$\text{AF } \frac{1}{\sqrt{2}} = 2,$$

AF = 2·83-tons Strut.

(ii) *Force Diagram*

The force diagram method is simply a combination into one single diagram of all the separate polygons obtained by the method of joints.

At the bottom of Fig. 16 the separate force polygons for the four joints have been combined into a single force diagram.

The space diagram is drawn to scale and lettered according to Bow's notation.

With parallel loads and reactions as in this case, the polygon of external forces for the frame is the straight line *abc* in which *cd* and *da* represent the support reactions.

The diagram is commenced with the right-hand reaction joint. From *c* on the force polygon draw *ce* parallel to CE on the space diagram to intersect a line drawn from *d* parallel to ED at point *e*. To determine if the members meeting at this joint are struts or ties the members are considered in a clockwise direction round the joint—i.e. in the order C-D-E. The force which member DE exerts on the right-hand joint is represented by vector *de* in the force polygon. The direction of *de* is from right to left on the *force diagram*, see Fig. 17. Therefore, the direction of *de* is *inwards* away from the joint, indicating that DE is in tension.

Now consider the next joint DAFE at the left-hand reaction. From *e* on the force polygon draw *ef* parallel to EF to intersect a line from '*a*' parallel to AF at the point *f*.

The polygon for this joint is now *dafe* in that order. The direction of vector *af* is downwards to the left, and, therefore, the arrowhead on member AF acts downwards *towards* the joint indicating a strut.

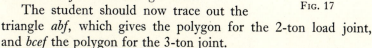

Fig. 17

The student should now trace out the triangle *abf*, which gives the polygon for the 2-ton load joint, and *bcef* the polygon for the 3-ton joint.

It should be observed that arrowheads are *omitted* from the force diagram. The reason for this can be seen from the two separate polygons *cde* and *dcfe* in Fig. 16.

The member ED features in both joints, and, therefore, the direction of the vector in the force diagram depends on which of the two joints is being considered—e.g. when considering the right-hand reaction joint the bottom member becomes DE, and when considering the left-hand one it is ED.

This method is completed by tabulating the magnitude and the nature of the force in each of the members.

(iii) *Method of Sections*

This method has the great advantage that the force in almost any member can be obtained by simply cutting the frame through that member and applying the principle of moments.

Consider the truss in Fig. 18 and let it be required to determine the forces in the three members cut by an imaginary cutting plane XX.

This cutting plane divides the truss into two parts.

Suppose the right-hand part of the frame is removed.

This will cause the left-hand part to collapse about the joint at R_1, as shown at (ii), unless *external* forces represented by BF, FE and ED are applied to the cut members as indicated. Now, in practice, the left-hand section of the truss is kept in equilibrium by the internal forces in the members BF, FE and ED. Therefore, the external forces applied in diagram (iii) actually represent the internal forces which we wish to calculate.

If we consider the equilibrium of the *left*-hand section of the truss, the external forces applied to the cut section must either push or pull at the joints on the *left* of the section XX. It is immaterial which 'sense' is initially assigned to these forces as the correct sense will be indicated by the sign of the final answer.

The left-hand section of the frame is now in equilibrium under the action of the reaction R_1, the 2-ton load and the three unknown

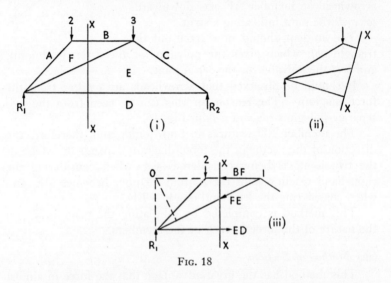

Fig. 18

forces BF, FE and ED. By taking moments of all the forces about the left-hand reaction—i.e. the point of intersection of the two unknown forces FE and ED—the value of BF can easily be found. Clearly, the balancing of forces about R_1 requires that the force in member BF should be directed to the left so that BF pushes on the cut section towards the 2-ton load joint. This indicates that member BF is a strut. Hence in this simple case the sense of the force can be obtained by inspection. Moments about point I will eliminate the forces BF and FE and give the force ED—its direction depending on the relative magnitude of the other two moments. Finally, moments about any point such as

O will give the value of the force in FE.

This method will now be applied to the given framework.

Force in BF

Cutting plane XX.   Take moments about R_1.
Assume BF is a strut, i.e. force
BF pushes on section XX.
Then $(2 \times 8) = BF \times 8$,
$$BF = 2 \text{ tons.}$$

Ans. is *Positive*.   Therefore,
assumption of a strut is correct.

BF is a Strut.

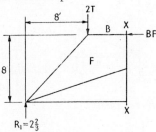

Force in ED

Cutting plane XX.   Take moments about point I.
Assume ED is a strut, i.e. force
ED pushes on cut section.
Then $(2\tfrac{2}{3} \times 18) + (ED \times 8)$
$$= (2 \times 10).$$
$$ED = -3 \cdot 5 \text{ tons.}$$

Ans. is *Negative*. Hence the
assumption that ED is a strut is
incorrect.

ED is a Tie.

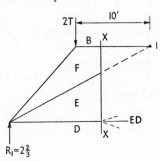

Force in FE

Cutting plane XX.   Take moments about point O.
Assume FE is a strut.
Then $FE.x. \quad + (2 \times 8) = ED \times 8$,
$$FE.8 \cos \theta + 16 \quad = 3 \cdot 5 \times 8.$$
$$FE = 1 \cdot 64$$
$$\text{tons.}$$

Ans. is *Positive*. Therefore, assumption of strut is correct.

FE is a Strut.

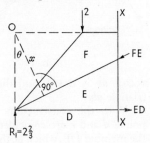

Force in AF

Cutting plane YY.   Take moments about point O.

Assume AF is a strut.

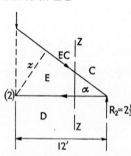

Then $AF.y + FE.x$ $= ED \times 8$.

$AF.8 \cos 45° + 1.64.8 \cos \theta = 3.5 \times 8$.

AF = 2.83

tons.

Ans. is *Positive*.   **AF is a Strut.**

(This result is more easily obtained by re-solution at the 2-ton load joint—see Method of Joints.)

Force in EC

Cutting plane ZZ.   Take moments about point (2).

Consider R.H. section of frame.

Assume EC is a strut.

Then $EC.z$ $= R_2 \times 12$.

$EC.12 \sin \alpha = 2\frac{1}{3} \times 12$.

EC = 4.2

tons.

Ans. is *Positive*.   **EC is a Strut.**

Example 4

The crane, shown Fig. 19, supports a vertical load of 2000 lb. by means of a chain.   Determine the forces in each member of the crane.

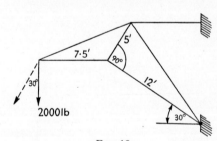

If the chain was pulled sideways to make an angle of 30° with the vertical, what effect would this have on the forces in the members?

Solution

A force diagram is shown opposite for the vertical load condition, Fig. 20 (i).   Com-mencing at the load joint, the

Fig. 19

vector *ab* is drawn and the forces in BC and CA obtained from the triangle *abc*.

With the construction of triangle *cda* it will be found that the points *d* and *e* are coincident.   This indicates that there is no force in the member DE, which is, therefore, redundant *for this particular loading*.

For the inclined load condition given in the question, the points *d* and *e* are not coincident as shown in the lower force diagram (ii). Hence for this loading DE is not a redundant member. The force diagram should be traced out round joint BEDC to verify that in this

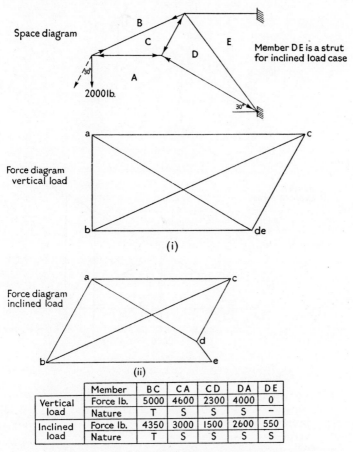

	Member	BC	CA	CD	DA	DE
Vertical load	Force lb.	5000	4600	2300	4000	0
	Nature	T	S	S	S	–
Inclined load	Force lb.	4350	3000	1500	2600	550
	Nature	T	S	S	S	S

FIG. 20

case DE is a strut. The effect of the inclined load on the other members is shown in the table accompanying the diagrams.

This example illustrates the practical necessity of including a member, such as DE, and stresses the importance of considering all the possible loading conditions which are likely to occur in a particular design.

Example 5

The cantilever frame, shown in Fig. 21, is hinged at D and supported by rollers at E. Determine, for the given loading, the forces in the members CD, BE and EF.

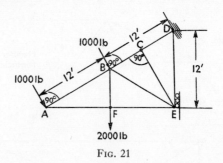

FIG. 21

Solution

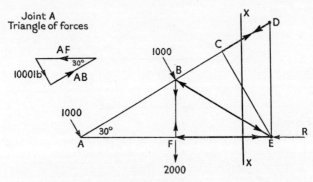

FIG. 22

Forces by Inspection

Joint F. For equilibrium at this joint member BF must supply 2000 lb. vertically upwards to balance the external load. (This should be apparent, since members AF and FE can provide no vertical component.)

Also for equilibrium Force in AF = Force in FE.

Joint C. By similar reasoning, member CE carries no force, i.e. CE is redundant.

Also Force in CD = Force in BC.

Force in CD (by Method of Sections). Fig. 22

Cutting Plane XX. Take moments about E. CD is a Tie (by Inspection).

$$CD \times 6\sqrt{3} = (1000 \times 18) + (1000 \times 6) + (2000 \times 6\sqrt{3}).$$
CD = 4320-lb. Tie.

Force in AF-FE (by Triangle Forces at Joint A)

$$AF = 2000\text{-lb. Strut.}$$
$$\therefore \ FE = 2000\text{-lb. Strut.}$$

Force in BE (by Resolution at Joint E)

For equilibrium at Joint E net horizontal force is zero.

To determine reaction R take moments of external forces about D. (The reaction at a roller support is at right-angles to the roller track.)

$$R \times 12 = (1000 \times 12) + (1000 \times 24) + (2000 \times 6\sqrt{3}),$$
$$R = 4730 \text{ lb.}$$

Then $\qquad$ R = Force in FE + Resolved part of force in BE.

$$4730 = 2000 + BE \cos 30.$$

BE = 3150-lb. Strut.

Example 6

The roof truss, shown in Fig. 23, is fixed at the right-hand support whilst the left-hand support rests on rollers. Determine, by calculation,

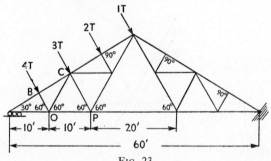

FIG. 23

the reactions at each support, both in magnitude and in direction, and also the force in each of the members BC, CO and OP. State whether each member is a strut or tie.

Solution

To determine Reactions.

The left-hand reaction will be vertical since the truss rests on rollers at this end. The resultant of the four external loads is 10 *tons*, and its line of action can be determined by taking moments about the left-hand support. Hence resultant load passes through C.

The truss can now be considered in equilibrium under the action of the 10-ton resultant load and the two reactions, as shown in Fig. 24.

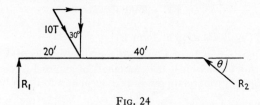

FIG. 24

Let the right-hand reaction R_2 make angle θ with the horizontal.

Then for vertical equilibrium $R_1 + R_2 \sin \theta = 10 \cos 30$,
and for horizontal equilibrium $R_2 \cos \theta \quad = 10 \sin 30$.

Solving these two equations gives

$$R_1 = R_2 \simeq \textbf{5·8 tons and } \theta = \textbf{30}°.$$

Consider Truss to left of section XX, Fig. 25.

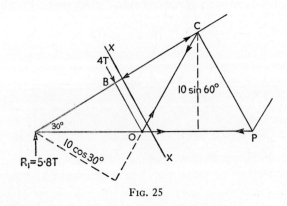

FIG. 25

Force in BC

Cutting plane XX. Take moments about O.
Assume BC is a Strut. Then BC $\times 5 = 5·8 \times 10$,

∴ **BC = 11·6-tons Strut.**

Force in CO

Cutting plane XX. Take moments about R_1.

Assume CO is a Tie.   Then $4 \times 10 \cos 30 = CO \times 10 \cos 30$,

$$\therefore CO = \textbf{4-tons Tie.}$$

Force in OP

Cutting plane XX.   Take moments about C.
Assume OP is a Tie. Then

$$OP \times 10 \sin 60 + 4 \times 10 \cos 30 = R_1 \times 15,$$

$$\therefore OP \simeq \textbf{6-tons Tie.}$$

Example 7

The diagram shows a simply supported truss.   Determine (*a*) graphically the forces, and nature of the forces, in the members (1), (2), (3) and (4), and (*b*) by calculation the forces, and nature of the forces, in members (5), (6) and (7).

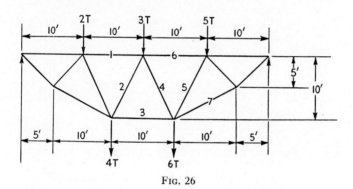

FIG. 26

Solution

Reactions

By moments about R_1,

$$40\ R_2 = (2 \times 10) + (4 \times 15) + (3 \times 20) + (6 \times 25) + (5 \times 30).$$

$$R_2 = \textbf{11 tons,} \qquad R_1 = \textbf{9 tons.}$$

(*a*) *Forces in members 1, 2, 3, 4 by Force Diagram*

The part of the force diagram required for these four members is shown in Fig. 27.   Note that the forces—commencing with *ab*—follow round the load line *ed* in the usual way.

(b) *Forces in members* 5, 6, 7 (*by Method of Sections*). See Fig. 28

Force in MN (5). Cutting plane XX. Take moments about point O.

Assume MN a Strut. Then MN . $x + (11 \times 5) = (5 \times 15)$.

MN = 1·49-ton Strut.

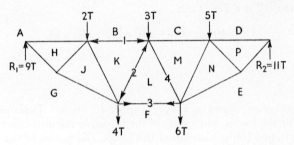

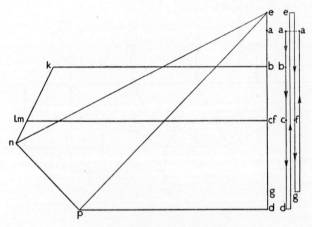

Member	1	2	3	4
Force ton	$12\frac{1}{2}$	$3\frac{1}{3}$	14	0
Nature	S	S	T	—

Fig. 27

Force in CM (6). Cutting plane XX. Take moments about point 4.
Assume CM a Strut. Then $CM \times 10 + (5 \times 5) = (11 \times 15)$.

CM = 14-ton Strut.

Force in NE (7). Cutting plane XX. Moments about point 2.
Assume NE a Tie. Then $NE \times y = (11 \times 10)$.

NE = 16·35-ton Tie.

Distance x. By similar triangles 012 and 245,

$$\frac{x}{02} = \frac{45}{2.4} \text{ or } \frac{x}{15} = \frac{10}{\sqrt{125}}$$

$$x = 6\sqrt{5}.$$

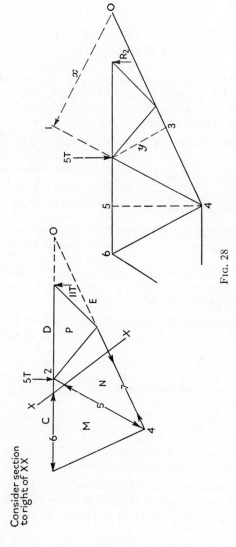

FIG. 28

Distance y. By similar triangles 023 and 456,

$$\frac{y}{02} = \frac{56}{46} \text{ or } \frac{y}{15} = \frac{5}{\sqrt{125}}$$

$$y = 3\sqrt{5}.$$

NB.—It is often simpler to measure distances such as x and y from a scale drawing.

EXERCISES I

1. A beam AB, 10 feet long, is simply supported at A and at a point C, 7 feet from A. Loads of 2, 3 and 1 tons are carried by the beam at distances of 3 feet, 5 feet and 10 feet, respectively, from A. Determine, graphically, the vertical reactions at A and C.

2. Determine, by means of the funicular polygon, the magnitude of the reaction supplied by the wall for the loaded beam shown in Fig. 29.

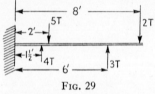

FIG. 29

3. A uniform bar AB, weighing 120 lb., is inclined at 20° to the horizontal plane and is supported by cords AC and BD. The cord BD is vertical and AC is inclined at 30° to the vertical, such that the angle CAB is 100° and DBA is 110°. Determine the horizontal force P, applied at B, which is required to keep the bar in this position if P is in the same vertical plane as the bar and the cords.

4. Fig. 30 shows a uniform horizontal beam AB, 18 feet long, which is hinged at B and freely supported at C, 6 feet from A. The beam weighs 4 tons and, in addition, carries loads of 8 tons and 24 tons at A and D, respectively, as shown. Determine the supporting reactions at B and C.   U.E.I.

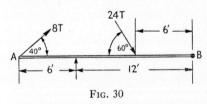

FIG. 30

5. (*a*) State the 'graphical' conditions that must obtain in order that a system of co-planar forces shall be in equilibrium.

(*b*) A horizontal beam XY, 20 feet long, is hinged freely at X and rests at Y on a smooth surface inclined at 45° to the horizontal. The beam carries vertical loads of 4 tons and 6 tons at points 6 feet and 14 feet, respectively, from X.

Draw the beam to scale and determine, graphically, the magnitude and direction of the reaction at X. Use the following scales : space 1 inch to 5 feet. Force 1 inch to 4 tons.   U.E.I.

6. A light beam, as shown in Fig. 31, carries a load of 50 lb. at D, and is supported in the horizontal position by three strings attached at A, B and C, the angles which the strings make with the beam being 90°, 60° and 45°, respectively. If the distances AD, DB and BC are each 10 feet, determine, graphically, the tensions in the three strings.

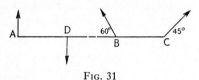

Fig. 31

7. Determine the resultant of the forces shown in Fig. 32 which act on a square of 2 feet side.

8. A uniform rod, whose length is 12 feet and weight 30 lb., is placed over a smooth peg so that one end rests against a smooth vertical wall. The distance of the peg from the wall is 9 inches. Find the position of equilibrium of the rod and the force between the rod and the peg.

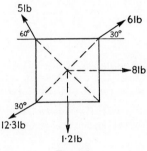

Fig. 32

9. A uniform beam AB weighs 100 lb. and is 12 feet long. It is hinged at A and supported in a horizontal position by a rope attached to a point C in the beam, 10 feet from A. The rope is inclined at 30° to the beam and is fastened to a wall vertically above A. The beam carries a vertical load of 80 lb. at the end B. Determine the hinge reaction at A and the tension in the rope.

10. The horizontal arm AB of a wall crane is 15 feet long and weighs 2 tons ; it carries a load of 4 tons at B, whilst the end A is pin-jointed to the wall. A strut is pin-jointed to the arm 5 feet from A and has its lower end pin-jointed to the wall at a point 6 feet from A.

Find graphically or otherwise :
  (a) the force acting in the strut.
  (b) the direction and magnitude of the wall reaction at A.

11. Fig. 33 shows a pin-jointed structure carrying vertical loads at the joints B, C and D and simply supported at the ends A and E.

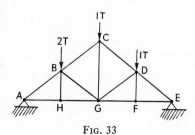

Fig. 33

CG = 15 feet and AH = HG = GF = FE = 10 feet.

Determine the reactions at the supports and draw the force polygon for the structure. Use scales of 1 inch = 10 feet and 1 inch = 1 ton. Write down the magnitude of the force in the member CG and determine which of the members are struts.   U.L.C.I.

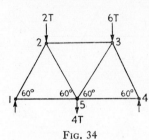

FIG. 34

12. Fig. 34 shows a pin-jointed frame carrying three vertical loads as shown.

(a) Determine 'by means of a funicular polygon' the supporting reactions at joints 1 and 4.

(b) Draw a force diagram for the frame and use it to determine the magnitude and nature of the forces in members 2-5 and 3-5.   U.E.I.

13. A workshop has a north-light roof, the dimensions of a typical bay being given in Fig. 35, which also shows the design loads in tons at the panel points. Points D and E are at the centre of their respective rafters.

(a) If the truss is pinned at A and B, calculate the reactions there.

(b) Draw the force diagram, hence determine the force in each

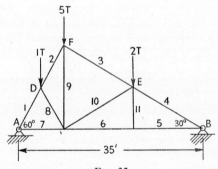

FIG. 35

member, writing its value on the bar in the space diagram. Distinguish between struts and ties by signs (+ compression, – tension). A table of forces is not required.

Suitable scales are : Linear 1 inch = 5 feet ; Force 1 inch = 1 ton.
I.Prod.E.

14. Due to a conveyor system, a roof truss carries loads of 3 tons and 2 tons as shown in Fig. 36.

Draw the force diagram for the truss and tabulate the magnitude and nature of the forces in the members.

Use the following scales: Space diagram 1 inch = 5 feet; Force diagram 1 inch = 1 ton.

I.Prod.E.

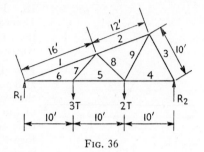

FIG. 36

15. A cantilever frame has horizontal upper and lower booms together with three other members which are inclined at 60° to the horizontal, as shown in Fig. 37.

When the structure supports a load of 2 tons at the end of the upper boom, find:

(*a*) the magnitude and direction of the reactions at the anchor pins;

(*b*) the forces in the members of the frame.

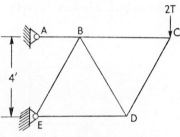

FIG. 37

Distinguish between struts and ties by signs (+ compression, – tension). Write the values found on the corresponding bar in the space diagrams. A table of forces is not required. Suitable scales are: Linear 1 inch = 2 feet; Force 1 inch = 1 ton.   I.Prod.E.

16. Fig. 38 represents a simply supported girder the joints of which may be regarded as pinned.

Determine the forces in the members 1, 2, 3 and 4, stating whether the members are struts or ties.

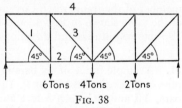

FIG. 38

17. Fig. 39 shows a pin-jointed structure simply supported at the ends and loaded as shown.

$$AJ = JI = IH = HG = 10 \text{ feet, and } BA = FG = 8 \text{ feet.}$$

Draw to a scale of 1 inch = 1 ton, the force diagram for the structure. Determine which members are struts. State the force in the member EI and the members in which the force is zero.

U.L.C.I.

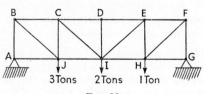

FIG. 39

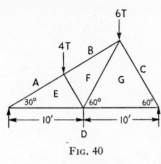

Fig. 40

18. Fig. 40 shows a loaded roof truss. Determine the values of the forces in the members BF, FG, GD, stating whether they are struts or ties.

19. The roof truss in Fig. 41 is hinged at B and supported on rollers at A. For the loading given, determine the values of the reactions at A and B in magnitude and direction. Also determine the forces in the members marked 1, 2, 3, stating whether they are struts or ties.

Use scales : 2 inches = 5 feet ; 1 inch = 1 ton.

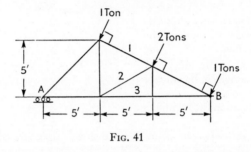

Fig. 41

20. Fig. 42 gives details of a loaded frame which is supported by a pin at Z and by a horizontal tie at joint Y. Determine the force in each member of the frame when carrying the loads indicated. Make a copy of the table shown in Fig. 42 and complete it with your results. State also the magnitudes and directions of the reactions at Y and Z.

N.C.T.E.C.

21. A simple frame of six bars is connected to a rigid vertical wall by pins in brackets at A and at E, 10 feet above A. There are two equal bars, AB and BC, each 5 feet long, and pinned at B, in a horizontal line from A. Two other equal bars, CD and DE, pinned at D, form a straight-line connection between C and E. The two remaining members connect D with A and with B. Vertical forces act downwards at D, C and B of respective magnitudes 1 ton, ½ ton and ¼ ton. By stress diagram, or otherwise, determine the forces acting on the bars, stating whether compressive or tensile, and find the value and direction of the reaction at A.   I.Mech.E.

22. The central panel ABCD of a pin-jointed truss is rectangular in form, 8 feet long by 6 feet deep. The bottom member is AD, with A on the left ; the top member is BC, with B on the left ; and there is a diagonal member AC. Loads of 18 tons and 9 tons act vertically downwards at A and D, respectively. The members of the adjacent panels act at the corners in such a way as to set up a horizontal force at A of 20 tons directed to the left; a horizontal force of 16 tons at D directed to the right; and an inclined force at B, the components of which are 15 tons vertically upwards and 20 tons horizontally to the right. Determine the nature and magnitude of the force in each member of the central panel and state the components of the force at C due to the connection with the adjacent panel. I.Mech.E.

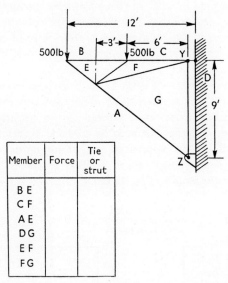

Member	Force	Tie or strut
B E		
C F		
A E		
D G		
E F		
F G		

FIG. 42

Chapter II

FRICTION

WE are concerned in work of this stage with the application of the principles of dry friction to machine elements involving screw threads, bearings and rotating discs, etc. A brief restatement of these principles is given as an introduction to this study.

5. Dry Friction.—Whenever motion is considered between surfaces which are assumed to be ideally smooth the *only* contact forces which can exist occur normal to the surface, Fig. 43 (i). For the case of actual surfaces in contact, a certain amount of roughness will always be present, so that any tendency for relative motion will cause tangential or friction forces to be developed which *oppose the motion*, Fig. 43 (ii).

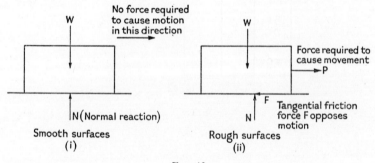

FIG. 43

The magnitude of the friction force which opposes motion will depend very much on the condition of the surfaces in contact, e.g. the degree of surface roughness and the extent of surface lubrication, and the following discussion is limited to the case of dry surface contact only.

6. Coefficent of Friction.—A block of weight W is shown in Fig. 44 (i) resting on a rough horizontal surface. If a horizontal force p is applied to the block which is insufficient to cause motion, it is evident that the surfaces have developed a friction force f which just balances p. The surface reaction, therefore, supplies *both* a normal force N equal to W and a tangential friction force f.

If p is gradually increased to the particular value P which is just sufficient to move the block, then, at the point of slipping, the

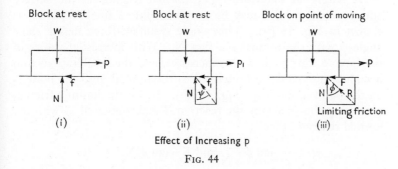

Effect of Increasing p

FIG. 44

friction force developed will have reached its maximum or limiting value F.   Fig. 44 (iii).

At this point, just prior to slipping, we have

$$\text{for horizontal equilibrium} \qquad P = F,$$
$$\text{and for vertical equilibrium} \qquad W = N.$$

It is found by experiment that the maximum friction force F which can be developed between a given pair of contacting surfaces is proportional to the *normal* reaction between the surfaces, N,

$$\text{i.e.} \qquad F = \mu N, \qquad . \qquad . \qquad . \qquad (i)$$

where μ is the coefficient of static friction.

The value of μ is very sensitive to even the slightest variation in surface conditions and because of this it is only possible to obtain average values by individual experiment.   It is well known that once a stationary body is put in motion the force required to keep it moving is slightly less than the maximum friction force to initiate the motion.   Friction between moving bodies is sometimes referred to as kinetic friction, and a coefficient of kinetic friction introduced in place of μ.   This implies that the friction coefficient depends on the magnitude of the relative velocity of the surfaces in contact. However, it is found that, providing the velocity is not excessive, sufficiently reliable calculations can be made on the basis of static friction without the refinement of kinetic coefficients. Values of μ may also be taken as independent of the areas of the surfaces in contact and of the magnitude of the normal reaction.

7. Angle of Friction.—The normal reaction at the surface and the friction force may be combined into a single reaction, as shown in Fig. 44 (ii). This single resultant force makes some angle ψ with the normal reaction N. With increasing values of the force p, this angle ψ also increases, and the resultant reaction becomes more and more inclined to the normal until the body is on the point of moving—Fig. 44 (iii). If, for the limiting friction conditions shown at (iii), the reaction R makes an angle ϕ with the normal reaction, then

$$\tan \phi = \frac{F}{N} \text{ and } F = \tan \phi.N.$$

Comparing this with equation (1), we have

$$\mu = \tan \phi \text{ or } \phi = \tan^{-1} \mu.$$

ϕ is called the angle of friction and is the angle between the resultant reaction of the plane on the block and the *normal to the surface of contact.*

8. Inclined Plane.—If in place of the horizontal surface in article 6 we substitute a rough *inclined* surface making an angle α with the horizontal, we have the condition of Fig. 45.

The force tending to cause the block to move down the plane is clearly the resolved part of W down the plane or W sin α, and if the body is at rest then the friction force—which opposes motion—must also equal W sin α, i.e. $f = $ W sin α. If, now, the angle of the plane is increased to some value θ, at which the body is just on the point of slipping, the condition of limiting friction will exist. Fig. 45 (ii).

For this case the component force causing motion will equal W sin θ, which must be equal to the maximum friction force F.

$$\therefore \ F = W \sin \theta.$$

The normal reaction between the contact surface at this angle
= W cos θ,

$$\therefore \; N = W \cos \theta.$$

Then since the coefficient of friction $= \dfrac{F}{\text{Normal Reaction}}$ *

we get $\qquad\qquad \mu = \dfrac{W \sin \theta}{W \cos \theta} = \tan \theta,$

Block at rest  Block on point of moving

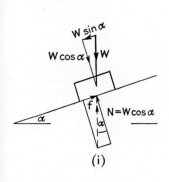

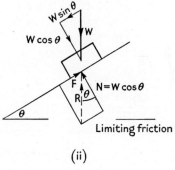

(i)   (ii)

FIG. 45

but in article 7 μ was shown to be equal to tan ϕ, where ϕ is the angle
of friction,

$$\therefore \; \tan \theta = \tan \phi,$$
$$\text{and} \quad \theta = \phi.$$

θ is called the angle of 'Repose'.

Example 8

A block of 50 lb. rests on a rough plane inclined at 30° to the
horizontal. $\mu = 0.3$.

Determine the force P acting parallel with the incline required to
cause the block :

        (a) to move up the plane,
        (b) to move down the plane.

* The normal reaction W cos θ must be used to calculate μ—not the force W.

Solution

Case (*a*). When motion takes place up the plane, the friction force F acts down the plane—Fig. 46 (a).

For equilibrium :

Resolving forces normal to the plane N = W cos 30°,

$$\therefore \ N = 50\frac{\sqrt{3}}{2}.$$

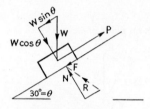

Motion up plane
F acts down the plane to
oppose motion
(a)

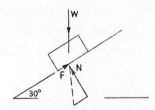

Motion down plane
F acts up the plane to
oppose motion
(b)

FIG. 46

Resolving forces parallel with the plane P = F + W sin 30°. .   (i)

Using the relationship $\mu = \dfrac{F}{N}$ and substituting for N, we get

$$F = \mu N = 0\cdot3 \times 50\frac{\sqrt{3}}{2},$$

$$\therefore \ F = \textbf{13 lb.}$$

Substituting this value of F in equation (i) gives P = 13 + 50 . ½,

$$\therefore \ P = \textbf{38 lb.}$$

Case (*b*). When the block is about to move down the plane, F will act up the plane.

The force causing the block to move down the plane is the weight component W sin 30°.

Equating forces parallel with the plane P + F = W sin 30°.

But F remains unchanged at 13 lb., ∴ P + 13 = 50 . ½,

$$\therefore \ P = \textbf{12 lb.}$$

Example 9

A casting weighing 0·5 ton is to be moved along a horizontal surface into its final machining position by means of a 12° wedge, as shown in

the diagram, Fig. 47. Neglecting the weight of the wedge, determine
the vertical force required on the wedge if the coefficient of friction
between the wedge and the casting surfaces is 0·3, and between the
supporting surfaces is 0·25.

Solution

A wedge is a particular case of an inclined plane (article 8) and can
be used for moving heavy loads through small distances. It is also
used as a locking device, e.g. cotter-joint, etc.

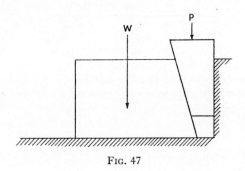

FIG. 47

The problem may be solved either by drawing or by calculation.

(a) *By Drawing*

Remembering that action and reaction are equal and opposite, we
may consider the equilibrium of the casting and the wedge separately.
As the wedge moves vertically downwards under the action of the
force P, the friction forces on the wedge must act upwards to oppose
the motion.

Hence the reactions R_1 and R_2 *on* the *wedge* act upwards at the
friction angles ϕ_1 and ϕ_2 to the normals marked *nn*.   The equal and
opposite reaction R_2 *on* the *casting* will, therefore, act downwards, as
shown in Fig. 48 (i).

The movement of the casting to the left will be opposed by the
friction between the casting and the horizontal supporting surface so
that the reaction R_3 will act towards the right at the angle ϕ_1 to the
normal.

Since the casting is in equilibrium under the action of the three
forces W_1, R_2 and R_3, the force diagram for the casting will give the
value of R_2.

With the value of R_2 thus obtained the force diagram for the wedge
may be drawn to give the value of the effort P.

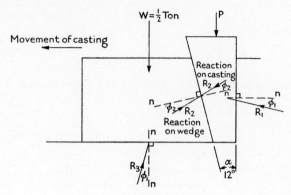

ϕ_1=Angle of friction for supporting surfaces $\phi_1=\tan^{-1}\frac{1}{4}=14°2'$
ϕ_2=Angle of friction between wedge and casting $\phi_2=\tan^{-1}0.3=16°42'$

(i)

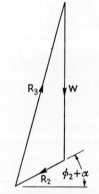

Force diagram for casting
(ii)

Force diagram for wedge
(iii)

P by measurement 0·119 ton

Fig. 48

(b) By Calculation.   Fig. 49

Consider the vertical and horizontal equilibrium of the casting and the wedge separately.

Casting

Since R_2 is inclined at ϕ_2 to the normal at the casting surface it will be inclined at the angle $(\alpha + \phi_2)$ to the horizontal.

For equilibrium :

Resolving forces horizontally :

$$R_2 \cos (\alpha + \phi_2) = R_3 \sin \phi_1 \qquad . \qquad . \qquad (1)$$

Resolving forces vertically :

$$W + R_2 \sin (\alpha + \phi_2) = R_3 \cos \phi_1. \qquad . \qquad (2)$$

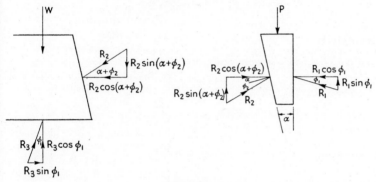

FIG. 49

Dividing equation (1) by equation (2), we get :

$$\frac{R_2 \cos (\alpha + \phi_2)}{W + R_2 \sin (\alpha + \phi_2)} = \tan \phi_1,$$

$$\frac{R_2 \cos (12° + 16° \ 42')}{W + R_2 \sin (12° + 16° \ 42')} = \tfrac{1}{4},$$

$$\therefore \quad \frac{R_2 \ 0.8771}{W + R_2 \ 0.4802} = \tfrac{1}{4},$$

$$\mathbf{R_2 = 0.33 \ W.}$$

Wedge

The wedge is in equilibrium under the action of the three forces P, R_1 and R_2.

For equilibrium :

Resolving forces horizontally :

$$R_2 \cos (\alpha + \phi_2) = R_1 \cos \phi_1. \qquad . \qquad . \qquad (3)$$

Resolving forces vertically :

$$R_2 \sin (\alpha + \phi_2) + R_1 \sin \phi_1 = P. \qquad . \qquad . \qquad (4)$$

Substituting the value of R_1 from equation (3) into equation (4) gives

$$P = R_2 \sin (\alpha + \phi_2) + R_2 \cos (\alpha + \phi_2) \tan \phi_1.$$

But $\qquad R_2 = 0.33W,$

$\qquad \therefore \ P = 0.33 \ W \times 0.4802 + 0.33 \ W \times 0.8771 \times \frac{1}{4},$

and $\qquad P = 0.231 \ W.$

For $W = \frac{1}{2}$ ton.

$$\mathbf{P = 0.115 \ ton.} \quad \text{(By Calculation.)}$$

Example 10

Fig. 50 shows a shaping machine mechanism. The upper end D of the slotted lever is pivoted to the ram and moves horizontally, while the lower end slides in a block C mounted in trunnion bearings. It is driven by the crank BA turning about the fixed centre B. CD = 26 inches, CB = 16 inches and BA = 5 inches. For the position shown $\theta = 60°$, find the turning moment at B to overcome a force of 100 lb. acting on the ram horizontally through the pivot D, assuming no friction.

If, now, friction at the three sliding contacts at C, A and D is to be considered, find the required turning moment if μ had a value of 0.25.

(I.Mech.E.)

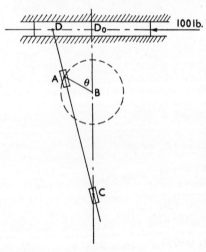

FIG. 50

Solution

The way in which the torque at B is used to overcome the 100-lb. force is illustrated in Fig. 51. This shows the clearances at the sliding surfaces greatly exaggerated and indicates how the forces are developed.

For the 'no friction' condition the only forces affecting the equilibrium of the *slotted lever* are R_1, R_2 and R_3, which act at right-angles
to the sliding surfaces.

The lines of action of the two forces R_2 and R_3 intersect at I, and,

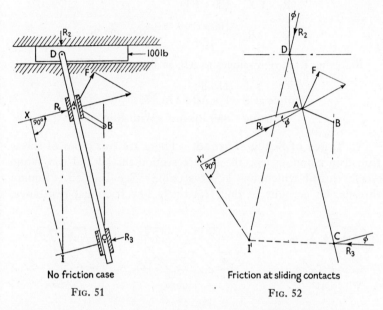

No friction case

Fig. 51

Friction at sliding contacts

Fig. 52

therefore, R_2 and R_3 will have no moment about this point.

Hence by moments about I, equilibrium requires

$$R_1 \times IX = 100 \times I.D.$$

$$R_1 = \frac{100 \times 27.45}{19.1} \quad \therefore \ R_1 = 144 \text{ lb.}$$

Resolving R_1 into components along AB and at right-angles to AB
gives

$$F = 100.4 \text{ lb.}$$
$$\therefore \text{ Torque at B} = F \times AB = 100.4 \times 5.$$

Torque = 502 lb. in. (No Friction.)

Friction at the sliding surfaces will cause the reactions R_1, R_2 and
R_3 to be inclined to the normal at these surfaces.

As the crank rotates in a clockwise direction, the lever will move
downwards in the sliding surfaces. The reactions R_1 and R_2 will,
therefore, be directed upwards to oppose this motion.

A.M.—D

R_2 will be inclined to the right of the normal at D in order to oppose the motion of the ram on its forward stroke.

From Fig. 52 the lines of action of R_2 and R_3 now intersect at I'. Then by moments

$$R_1 \times I'X' = 100 \times I'D,$$

$$R_1 = \frac{100 \times 25 \cdot 9}{12 \cdot 5} \quad \therefore \ R_1 = 207 \ lb.$$

Resolving R_1 at right-angles to AB, as above, gives

$$F = 176 \ lb.$$
$$\text{Torque at } B = F \times AB = 176 \times 5.$$
$$\textbf{Torque} = \textbf{880 lb. in.} \quad \text{(With Friction.)}$$

9. Types of Screw Thread.—There are two types of screw thread in common use, the square section thread and the Whitworth thread which has a vee-section. See Fig. 53. Square threads, on account of their relatively low frictional resistance,

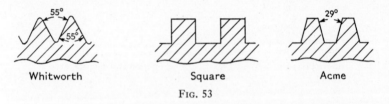

Whitworth   Square   Acme

FIG. 53

are most used for transmitting power, whereas the higher frictional resistance offered by the vee-thread makes it more suitable as a fastening device. Of the two, vee-threads are the stronger, although, as the thread face is inclined to the line of thrust, there is a force component which tends to burst the nut—see Fig. 57 (c). The advantages of both these threads are combined in the Acme type thread.

10. Square Thread.—Consider the square-threaded screw in the fixed nut, Fig. 54 (i), which supports an axial load W.

If one turn of the screw is unwrapped, as in Fig. 54 (ii), the helix angle of the thread is given by $\alpha = \tan^{-1} \dfrac{L}{2\pi r}$, where L is the lead and $2\pi r$ the mean circumference.

The screw can be imagined as made up of a number of small rectangular elements, as illustrated in Fig. 54 (ii). These elements rest on the inclined surface of the nut in the same way as the block on the inclined plane in article 8—see Fig. 45. We can, therefore, consider the rotation of a screw as similar to the movement of a block up an inclined plane, although, in the case of a thread, the force P is applied horizontally and not parallel with the incline, cf. Fig. 46.

For each individual element the reaction at the thread surface

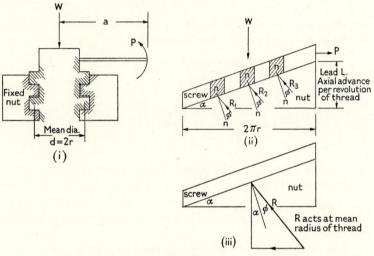

Fig. 54

will be inclined at the friction angle ϕ to the normal nn. The sum of all the elemental reactions R_1, R_2, etc., may be combined into a single reaction R, which is assumed to act at the mean radius of the thread.

The horizontal component of the reaction force R·

$$= R \sin (\alpha + \phi).$$

This force acting at the mean thread radius gives the mean torque $= R \sin (\alpha + \phi) \times r$.

If the horizontal force P, applied at a radius 'a', is just sufficient to move the screw up the inclined thread of the nut, then the applied torque $= Pa$.

For horizontal equilibrium,

$$Pa = Rr \sin (\alpha + \phi). \qquad . \qquad . \qquad . \qquad (1)$$

For vertical equilibrium the total thrust W must be balanced by the vertical component of R,

$$\text{i.e. } W = R \cos (\alpha + \phi). \qquad \qquad (2)$$

Dividing equation (1) by equation (2), we get

$$\frac{Pa}{W} = \frac{Rr \sin (\alpha + \phi)}{R \cos (\alpha + \phi)},$$

or
$$Pa = Wr \tan (\alpha + \phi). \qquad \qquad (3)$$

An alternative expression for this torque in terms of μ and the thread dimensions can be obtained since

$$\tan (\alpha + \phi) = \frac{\tan \alpha + \tan \phi}{1 - \tan \alpha \tan \phi} \; ; \; \tan \phi = \mu \; ; \; \tan \alpha = \frac{L}{\pi d}.$$

Hence
$$Pa = Wr \left\{ \frac{\dfrac{L}{\pi d} + \mu}{1 - \dfrac{L}{\pi d} \mu} \right\} = Wr \left\{ \frac{L + \pi d \mu}{\pi d - \mu L} \right\}.$$

NB.—For a single-start thread, Lead = pitch p.

Equation (3) gives the actual torque required to move the screw when limiting friction conditions exist. For the ideal case of no friction ϕ is zero, and then the ideal torque becomes $Wr \tan \alpha$.

The efficiency of the thread may then be expressed as

$$\text{Efficiency } \eta = \frac{\text{Ideal Torque}}{\text{Actual Torque}} = \frac{Wr \tan \alpha}{Wr \tan (\alpha + \phi)},$$

or
$$\eta = \frac{\tan \alpha}{\tan (\alpha + \phi)}.$$

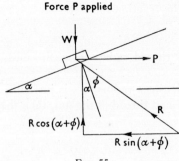

Force P applied

FIG. 55

If the applied force P in Fig. 55 is removed the only force resisting the tendency for the screw to slide *backwards* in the nut will be that provided by friction. The friction angle ϕ must then be drawn on the *other side* of the normal to the thread surface in the direction opposing this motion. There are then three possibilities which must be considered, viz. (*a*) when $\phi > \alpha$; (*b*) $\phi = \alpha$; (*c*) $\phi < \alpha$. Fig 56 illustrates these three cases and indicates that when $\phi < \alpha$, a torque equal to Wr tan ($\alpha - \phi$), must be supplied to prevent the screw moving backwards. Cf. equation 3.

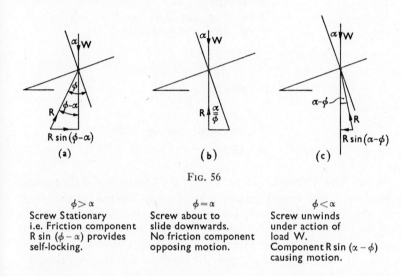

FIG. 56

$\phi > \alpha$	$\phi = \alpha$	$\phi < \alpha$
Screw Stationary i.e. Friction component R sin ($\phi - \alpha$) provides self-locking.	Screw about to slide downwards. No friction component opposing motion.	Screw unwinds under action of load W. Component R sin ($\alpha - \phi$) causing motion.

Example 11

A load of 0·5 ton is supported by a screw-jack having a single start square thread. The mean diameter of the thread is 2 inches and the coefficient of friction is 0·3. Determine the pitch of the screw so that it will not unwind by itself and also the horizontal force required at the end of a lever 15 inches long to raise the load. What will be the efficiency of the screw ?

Solution

If the screw is not to unwind itself α must not be greater than ϕ, i.e. $\alpha \not> \phi$, and therefore the limiting condition will be when $\alpha = \phi$. See Fig. 56 (b).

Since $\qquad \mu = \tan \phi = 0\cdot3.$
For this case $\qquad \tan \phi = 0\cdot3 = \tan \alpha \quad \therefore \; \alpha = \tan^{-1} 0\cdot3$
$$\alpha = 16\cdot7°.$$

Also for a single start thread

$$\tan \alpha = \frac{p}{\pi d},$$

$$\therefore \ 0{\cdot}3 = \frac{p}{2\,.\,\pi}. \quad \therefore \ p = 2\pi \times 0{\cdot}3.$$

$$p = 1{\cdot}88''.$$

Torque $\qquad\qquad Pa = Wr \tan{(\alpha + \phi)}.$

Then $\qquad\qquad P \times 15 = 1120\,.\,1\,.\,\tan 33{\cdot}4°.$

Since $\qquad\qquad \alpha = \phi = 16{\cdot}7°$ and $r = 1$ inch.

$$\therefore \ P = \frac{1120 \times 0{\cdot}66}{15}.$$

$$P = 49{\cdot}3 \ \textbf{lb.}$$

Efficiency $\qquad\qquad \eta = \dfrac{\tan \alpha}{\tan{(\alpha + \phi)}} = \dfrac{\tan \phi}{\tan 2\phi} = \dfrac{0{\cdot}3}{0{\cdot}66},$

$$\therefore \ \eta = \textbf{45·5\%.}$$

11. Vee Thread.—In the case of the square thread the normal reaction at the thread surface is inclined to the axis of the screw in one plane only—i.e. in the plane YOZ, due to the helix angle of the thread—Fig. 57 (a) and (b). For the vee thread the reaction is inclined to the screw axis in two planes, as shown in Fig. 57 (c) and (d), and this complicates the analysis for the vee thread.* Since α is small an acceptable treatment for this thread is to assume that α is zero. This means that R in Fig. 57 (c) is taken as the true normal reaction.

Then if $\qquad\qquad$ W = Axial Load,

the normal reaction R is increased to $W/\cos \beta$. $\ \beta$ = semi-angle of thread.

But Friction Force F = $\mu \times$ normal reaction,

$$\therefore \ F = \mu \ W/\cos \beta,$$

or $\qquad\qquad F = W\mu^1$, where $\mu^1 = \dfrac{\mu}{\cos \beta} = \mu \sec \beta.$

μ^1 can, therefore, be used in place of μ in equation (3) for the square thread.

Example 12

The mean diameter of a single-start thread of $\frac{1}{8}$ inch pitch is $0{\cdot}8$ inch and the angle of the thread is $55°$. Determine the force required

---

* See *Theory Machines*, B. B. Low (Longmans).

at the end of a spanner, 12 inches long, to produce a tension of 1000 lb. in the bolt when the nut is tightened. Friction at the bearing surface of the nut may be assumed concentrated at a mean radius of 0·9 inch. Coefficient of Friction for screw and bearing surface 0·2.

Solution

The torque at the spanner is required not only to overcome the

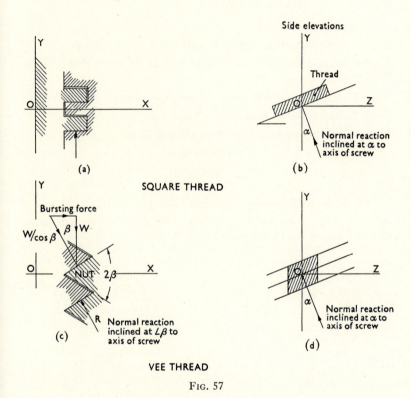

Side elevations

(a)

SQUARE THREAD

(b)

(c)

VEE THREAD

Fig. 57

(d)

friction of the screw thread but also the friction between the nut and its bearing surface.

Friction torque at screw :

μ^1 for vee thread $=\mu/\cos \beta$.

Then
$$\mu^1 = \frac{0\cdot2}{\cos 22\cdot5°} = \frac{0\cdot2}{0\cdot924} = 0\cdot217,$$

$$\alpha = \tan^{-1} \frac{\text{pitch}}{\pi d} = \tan^{-1} \frac{\frac{1}{8}}{\pi \times 0.8} \quad \therefore \ \alpha = 2° \ 51',$$

$$\phi = \tan^{-1} \mu^1 = \tan^{-1} 0.217 \quad \therefore \ \phi = 12° \ 15'.$$

Torque $= Wr \tan(\alpha + \phi)$,

$$\therefore \ \text{Torque} = 1000 \times 0.4 \tan 15° \ 6',$$

where 0·4″ is the mean radius.

Then   Torque $= 108$ lb. in.

Friction torque at nut :

Friction force at bearing surface $F = \mu W = 0.2 \times 1000 = 200$ lb.

This force is assumed to be concentrated at the mean radius of the bearing surface.

Then Friction Torque FR $= 200 \times 0.9 = 180$ lb. in.

Total Torque supplied by spanner $= 108 + 180 = 288$ lb. in.

Then if P is the required force at a radius of 12 inches,

$$P \times 12 = 288.$$

P = 24 lb.

12. Journal Bearing.—Journal bearings are used to support lateral loads. The bearing in Fig. 58 supports the journal or shaft of radius r under a lateral load W.

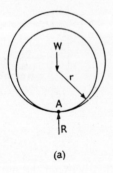

(a)

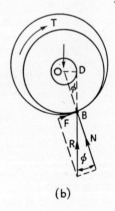

(b)

FIG. 58

When the shaft is at rest contact between shaft and bearing will be as shown at (a) where the clearance between the *dry* surfaces has been exaggerated. The bearing reaction at A is equal to the load W and is in line with it. When the shaft begins to rotate it rolls up the bearing surface in a direction *contrary* to the shaft rotation until limiting friction conditions occur at some point B. Slip now occurs and the shaft then stays in this position during rotation.*

Since friction opposes the rotation of the shaft the bearing surface reaction R will be inclined at angle ϕ to the normal OB at the point of contact B.

R is the resultant of the friction force F and the normal reaction N.

Vertical equilibrium requires that R = W, although, due to the rotation, the forces will not be in line.

For rotational equilibrium, these forces must form a couple equal and opposite to the applied couple T required to maintain the rotation.

i.e.   T = W. OD = frictional torque.

The reaction R is tangential to the small circle centre O and radius OD, which is called the friction circle.

$$\therefore \ T = W. \ OD = Wr \sin \phi.$$

Since for small angles sin ϕ is approximately equal to tan ϕ, we may write
$$T = Wr \tan \phi$$
or
$$T = \mu \ Wr.$$

This equation gives the maximum value of the torque required to overcome friction between *dry* surfaces.

If N is the speed of rotation in RPM,

$$\text{Frictional HP} = \frac{2\pi N.\mu Wr}{33000}.$$

13. Disc Friction.—Disc friction occurs when surfaces having relative motion are brought together by means of an axial thrust. For example, in thrust bearings which are used to support shafts under axial load, power is lost due to disc friction, whereas, in the case of clutch plates, disc friction is used as a means of transmitting power from one shaft to another.

* For lubricated bearings, the shaft moves round the bearing in the same direction as the shaft rotation (for later studies, see *Theory of Machines*, Bevan, Longmans).

In Fig. 59 the thrust pad A is supporting an axial load W by means of a rotating ring B having inner and outer radii R_1 and R_2, respectively.

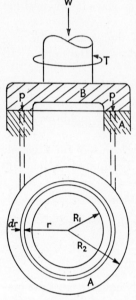

It is required to determine an expression for the torque necessary to overcome the frictional resistance to the rotation of the shaft. If p is the normal pressure at any elemental ring of bearing surface radius r and thickness dr, then axial load supported by this elemental ring

$$= p.2\pi r dr \quad (1) \text{ (pressure} \times \text{area).}$$

The tangential friction force

$$= \mu p 2\pi r dr \quad (\mu \times \text{Normal load).}$$

Moment of this friction force about shaft axis

$$= \mu p 2\pi r dr . r. \quad . \quad (2)$$

Then total frictional moment T

$$= \int_{R_1}^{R_2} \mu p 2\pi r^2 dr. . \quad (3)$$

FIG. 59

Equation (3) gives the torque and hence the horse power lost in friction for a thrust bearing, and in the case of a friction clutch the maximum horse power which can be transmitted under limiting friction conditions.

If μ is assumed constant at all points of the surface, this equation becomes

$$T = 2\pi\mu \int_{R_1}^{R_2} p r^2 dr.$$

(Constant terms may be taken outside the integral sign.)

A further simplification can be made if it is assumed that the pressure p is uniformly distributed across the bearing surface.

With p constant we get

$$T = 2\pi\mu p \int_{R_1}^{R_2} r^2 dr. = 2\pi\mu p \left[\frac{r^3}{3} \right]_{R_1}^{R_2}$$

$$\therefore \quad T = \tfrac{2}{3}\mu p(R_2{}^3 - R_1{}^3).$$

But $p = \dfrac{\text{Load}}{\text{Area}} = \dfrac{W}{\pi(R_2^2 - R_1^2)},$

$$\therefore\ T = \tfrac{2}{3}\mu W \cdot \dfrac{R_2^3 - R_1^3}{R_2^2 - R_1^2}. \qquad\qquad (4)$$

For a solid disc of radius R,

$$R_1 = 0,\ R_2 = R.$$

Then $\qquad\qquad T = \tfrac{2}{3}\mu WR.$

For new surfaces which may be considered to make perfect contact at all points, the simple assumption of uniform axial pressure is justified. However, after some wear of the bearing surfaces, it is found that the frictional torque becomes less than that calculated by equation (4).

It is reasonable to suppose that the rate of wearing between two surfaces in contact will depend in some way on the pressure between the surfaces and on their relative speed, i.e. rate of wear $\propto pv$, where v is the linear speed. As linear speed is equal to ωr, v is proportional to r. Therefore, the rate of wear is proportional to the product pr (since ω is constant). If the rate of wear is assumed uniform then $pr = C$, where C is a constant.

Then from equation (1)

Axial load supported
   by element $\qquad = p2\pi r dr,$

and substituting $C = pr$
   we get $\qquad\qquad 2\pi C dr.$

Total load $\qquad W = 2\pi C \displaystyle\int_{R_1}^{R_2} dr,$

and $\qquad\qquad W = 2\pi C(R_2 - R_1).$

From equation (2)

Moment of friction force
   about shaft axis $\quad = 2\pi\mu C r dr.$

Total Moment $\qquad T = 2\pi\mu C \displaystyle\int_{R_1}^{R_2} r dr.$

Then $\qquad\qquad T = \pi\mu C(R_2^2 - R_1^2).$

Substituting for $\quad C = \dfrac{W}{2\pi(R_2 - R_1)}$ from above,

we get $\qquad T = \dfrac{\mu W}{2} \cdot \dfrac{R_2{}^2 - R_1{}^2}{R_2 - R_1},$

$$= \frac{\mu W}{2} \cdot \frac{(R_2 + R_1)(R_2 - R_1)}{R_2 - R_1},$$

and hence $\qquad T = \dfrac{\mu W}{2}(R_1 + R_2).$

For a solid disc of
  radius R, $\qquad T = \dfrac{\mu W R}{2}.$

Example 13

What power can be transmitted by a pair of disc plates having inside diameter 3 inches and outside diameter 6 inches, when running at 1500 rev./min.? The axial load is constant at 750 lb. $\mu = 0.3$.

Solution

When the plates are new uniform pressure may be assumed.

$$\therefore\ T = \tfrac{2}{3}\mu W \cdot \frac{R_2{}^3 - R_1{}^3}{R_2{}^2 - R_1{}^2},$$

$$T = \tfrac{2}{3} \times 0.3 \times 750 \cdot \frac{216 - 27}{36 - 9}.$$

$$T = 1050 \text{ lb. in.} = 87.5 \text{ lb. ft.}$$

Then $\qquad HP = \dfrac{2\pi NT}{33000} = \dfrac{2\pi \times 1500 \times 87.5}{33000}.$

HP = 25.

After a running-in period uniform wear conditions are more likely to exist.

Then $\qquad T = \dfrac{\mu W}{2}(R_2 + R_1),$

$$T = \frac{0.3 \times 750}{2}(6 + 3),$$

$$T = 1012 \text{ lb. in.} = 84.3 \text{ lb. ft.}$$

Then $\qquad$ **HP = 24.1.**

Note that the uniform wear condition gives the lower horse power for a constant axial load. Therefore, the actual power which can be transmitted is 24·1 horse power.

Example 14

A vertical two-start square-threaded screw 4·375 inches external diameter and 0·75 inch pitch supports a vertical load of 3000 lb. The nut for the screw is fitted in the hub of a gear wheel having 80 teeth which gears with a pinion of 60 teeth. The mechanical efficiency of the wheel and pinion is 85 per cent. The axial thrust of the vertical screw is taken on a collar bearing 4 inches inside diameter and 10 inches outside diameter, for which a 'uniform pressure' condition may be assumed. If the coefficient of friction for the vertical screw is 0·15 and for the collar is 0·20, determine the torque required on the pinion shaft in order to raise the load.

Solution

Torque for screw.

Depth of square thread $= \dfrac{\text{Pitch}}{2} = 0.375''$,

$\therefore$ Mean thread diameter $= 4.375 - 0.375 = 4.0''$.

$$\text{Tan } \alpha = \frac{\text{Lead}}{\pi d} = \frac{2 \times \text{pitch}}{\pi d},$$
(Double-start thread.   L $=2p$.)

$$\therefore \text{ Tan } \alpha = \frac{2 \times 0.75}{\pi \cdot 4} = 0.1194.$$

Hence $\qquad\qquad \alpha = 6° 49'.$

Friction angle $\phi = \tan^{-1} 0.15 = 8° 32'.$

Torque on screw $Pa = W\dfrac{d}{2} \tan (\alpha + \phi)$. $d =$ mean thread diameter,

$\qquad \therefore$ Torque $= 3000 \times 2 \times \tan 15° 21'.$

$\qquad$ Torque $= 1650$ lb. in.

Torque for collar bearing.

$$\text{Torque} = \tfrac{2}{3}\mu W \frac{R_2{}^3 - R_1{}^3}{R_2{}^2 - R_1{}^2}.$$

Then $\qquad\qquad$ Torque $= \tfrac{2}{3} \cdot 0.2 \cdot 3000 \cdot \dfrac{(1000 - 64)}{(100 - 16)},$

$\qquad\qquad$ Torque $=$ lb. 4450 in.

Therefore the total torque required at the gear wheel

$$= 1650 + 4450 = 6100 \text{ lb. in.}$$

Now the mean radius of the pinion is less than that of the gear wheel, and since torque is proportional to radius the torque on the pinion will be less than 6100 lb. in. But the radius of a gear wheel is proportional to its number of teeth.

Therefore Torque is proportional to the number of gear teeth.

Hence Torque at Pinion
(for 100% efficiency) $= 6100 \times \dfrac{60}{80}$.   Pinion, 60 teeth,
                                                wheel, 80 teeth.

$$= 4575 \text{ lb. in.}$$

Actual Torque required with
85% gearing efficiency $= 4575 . \times \dfrac{100}{85}$,

∴ **Torque at Pinion** $= 5382$ **lb. in.**

EXERCISES II

1. Find the least value of a force P required to pull a block weighing 20 lb. along a rough horizontal plane, if (a) P acts horizontally, (b) P acts 30° to the horizontal. $\mu = 0.5$.

2. A body, weighing 20 lb., is found to slide down a plane with uniform velocity when the angle of inclination of the plane is 30°. If this body is pulled up the plane a distance of 10 feet, how much work is done ?

3. The inclination of a plane to the horizontal is 27°. What force at an angle of 20° to the plane will just move a body weighing 100 lb. up the plane if the coefficient of friction is 0·2 ?

What is the inclination of the minimum force required to move the body up the plane ?

4. A body, of weight w lb., on a plane inclined at 20° to the horizontal, and for which the coefficient of friction is μ, is acted upon by a force applied upwards and parallel to the plane. When this force has a value of 12 lb. the body slides steadily downwards ; when the value is 35 lb. the body moves steadily upwards. Deduce from these results the values of w and μ.

A different body, of weight 100 lb., and with a surface for which, on the same plane, the friction coefficient is 0·15, is to be moved by a force P, directed at an angle of 15° to the plane, i.e. at 35° to the hori-

zontal. Calculate the value of P which will cause steady upward movement ; and also the value to which P must be reduced before downward
movement becomes possible. Any formulae used should be established
or explained by vector diagrams of forces. I.Mech.E.

5. A plane has an incline of 1 in 8 (measured along the incline). It
is found that a force of 70 lb. acting down the incline and parallel to it
is just sufficient to move a block of stone weighing 2 cwt. Find by
drawing or calculation :

(a) the force acting up the plane at an angle of 20° above it which
would just move the block;

(b) the incline down which the block would just slide.

6. A block weighing 5 tons is raised by
means of the wedge shown in Fig. 60.
If the taper of the wedge is 1 in 10, and
the coefficient of friction between all contacting surfaces is 0·2, determine the value
of the force P required.

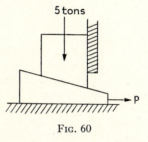

FIG. 60

7. The mechanism for a special lifting
table is shown in Fig. 61. When the hand
wheel is turned clockwise, the screw is
drawn to the right, lifting the table. Taking
account of screw and collar friction only, calculate the force (in lb.)
required at the hand wheel to just lift a load of $\frac{1}{2}$ ton when the toggle

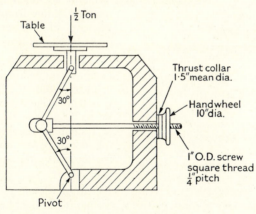

FIG. 61

links, which are of equal length, make an angle of 30° to the vertical.
The coefficient of friction for the screw and collar are 0·1 and 0·15,
respectively. I.Prod.E.

8. Define 'coefficient of friction' and 'angle of friction'. If a horizontal force P moves a weight W upwards on a plane inclined to the horizontal at α show that P = W tan $(\alpha + \phi)$, where ϕ is the angle of friction.

The rotation of a square-threaded screw on a machine moves a nut and thereby gives steady translational motion to a slide at a speed of 5 ft./min. The screw is 2 inches mean diameter and $\frac{1}{2}$ inch pitch; and the coefficient of friction between nut and screw is 0·1. The axial load on the screw is taken by a ball thrust bearing, the frictional effect of which may be neglected. If the total resistance to the motion of the slide is 200 lb., determine the power, in ft.lb. per min., used in the operation.   I.Mech.E.

9. A body on a plane inclined at α to the horizontal is moved upwards by a horizontal force. Show that the efficiency of the lift is given by the ratio of tan α to tan $(\alpha + \phi)$, where ϕ is the angle of friction.

A screw-jack is used horizontally in sliding a bedplate into position on its foundation. The bedplate weighs 4 tons and the coefficient of friction between it and the foundation is 0·25. The screw of the jack has a mean diameter of 2 inches and a pitch of $\frac{1}{2}$ inch; the coefficient of friction is 0·1. The axial thrust is carried on a collar of mean diameter 2·7 inches, for which the coefficient of friction is 0·15. Determine the torque required on the jack and the efficiency of the operation.   I.Mech.E.

10. If an anticlockwise torque of 200 lb. ft. is applied to the brake drum in Fig. 62, determine the least value of W which will prevent rotation of the drum. The co-efficient of friction between brake block and drum is 0·25.

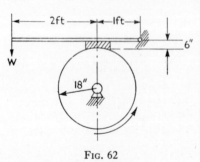

FIG. 62

11. A rotor weighing 8 tons is mounted in bearings 6 inches diameter, and the torque required to make it revolve is 45 lb. ft. Determine the coefficient of friction at the bearings and the horse power absorbed by friction when the speed is 300 rev./min.

12. What is meant by the term 'coefficient of friction'. A shaft 12 inches in diameter is supported in a bearing, the normal load between shaft and bearing being 12 tons. When the shaft turns at 85 rev./min., 15 horse power is used in overcoming friction.

Find (*a*) the frictional torque in lb. ft.,
      (*b*) the frictional resistance at the bearing,
      (*c*) the coefficient of friction.

Chapter III

DYNAMICS I—KINEMATICS

KINEMATICS is that part of dynamics which deals with the motion of bodies without considering the forces required to produce the motion.

This chapter, then, is concerned with *moving* bodies and the determination of their velocities and accelerations.

14. Revision.—*Velocity* of a body is its rate of change of displacement with respect to time.

Velocity is a vector quantity and hence it defines direction as well as magnitude (speed).

Acceleration of a body is the rate of change of its velocity with respect to time.

Acceleration is also a vector quantity since it involves both rate of change of direction as well as rate of change of speed.

Acceleration may or may not be uniform and it may occur both in the direction of motion and at right-angles to it.   See article 17.

For uniformly accelerated motion only we have the following equations :

Linear Motion	Angular Motion
$s = vt$	$\theta = \omega t$
$s = \left(\dfrac{u+v}{2}\right)t$	$\theta = \left(\dfrac{\omega_1 + \omega_2}{2}\right)t$
$v = u + ft$	$\omega_2 = \omega_1 + \alpha t$
$s = ut + \frac{1}{2}ft^2$	$\theta = \omega_1 t + \frac{1}{2}\alpha t^2$
$v^2 = u^2 + 2fs$	$\omega_2{}^2 = \omega^2{}_1 + 2\alpha\theta.$

Notation :

Displacement	s	θ
Initial Velocity	u	ω_1
Final Velocity	v	ω_2
Acceleration	f	α
Time taken	t	t
(or time at any instant)		

The symbols v and ω are also used to indicate the velocity at any instant.

For rising and falling bodies the effect of gravity is to produce an acceleration of g ft./sec.^{2} towards the earth, and in such cases f can be replaced by g in the above equations.

Example 15

A body moves over 15 feet in the 1st second of its motion, over 63 feet in the 3rd and over 135 feet in the 6th second. Show that these distances are consistent with the supposition of uniform acceleration.

Solution

From the information given for the first 3 seconds we can assume uniform acceleration and calculate f. This value can then be used to evaluate the distance travelled in the 6th second for a comparison with the actual distance given—see Fig. 63.

Let $u =$ initial velocity at $t = 0$.
             $v_1 =$ velocity reached at end of 1 second.
             $v_2 =$ velocity reached at end of 2nd second, etc.
             $s_2 =$ distance travelled in 2nd second.

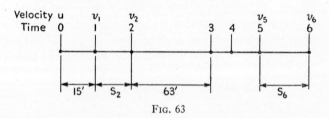

FIG. 63

During the 1st second (i.e. when $t = 1$), the distance travelled can be calculated from

$$s = ut + \tfrac{1}{2}ft^2.$$

Hence for $t = 1$, $15 = u + \tfrac{1}{2}f.$.   .   .   (1)
Similarly for $t = 2$, $15 + s_2 = u \cdot 2 + \tfrac{1}{2} \cdot f \cdot 4.$.   .   (2)
and for $t = 3$, $15 + s_2 + 63 = u \cdot 3 + \tfrac{1}{2}f \cdot 9.$.   .   (3)

These three equations give three unknowns—viz. s_2, u and f.

Eliminating u. from (1) and (2) we get $s_2 = 15 + f$.   (4)
Eliminating u from (1) and (3) $s_2 = 3f - 33.$
Then eliminating s_2 we get $f = \textbf{24 ft./sec.}^2$
From equation (4)—using this value of f— $s_2 = \textbf{39 ft.}$
and finally $u = \textbf{3 ft./sec.}$

Now velocity at *beginning* of 6th second (equals velocity at end of 5th second) is given by $v = u + ft$,

so that velocity at beginning of 6th second $= u + f5$,

and velocity at end of 6th second $\quad = u + f6$.

$\therefore$ Average velocity during 6th second $\quad = \dfrac{u + 5f + u + 6f}{2}$

$$= u + \frac{11}{2}f.$$

Distance travelled in 6th second = Average velocity $\times$ time,

$$\text{i.e. } s_6 = \left(u + \frac{11}{2}f\right) \times 1.$$

Assuming uniform acceleration $f = 24$ and $u = 3$.

Then $\qquad\qquad\qquad s_6 = \left(3 + \dfrac{11}{2} \times 24\right) = \textbf{135 ft.}$

This is the same as the distance given for the 6th second.

Hence distances are consistent with uniform acceleration.

Example 16

A body is projected upwards from a point on level ground with a velocity of 100 ft./sec. at an angle of 30° to the horizontal. Working from the expressions for uniformly accelerated motion in a straight line, determine (*a*) the time of flight, (*b*) the distance from the point of projection to the point where the body strikes the ground, (*c*) the maximum height reached.   (N.C.T.E.C.)

Solution

Let u = velocity of projection.

This velocity can be resolved into two components at right-angles: $u \cos \theta$ horizontally and $u \sin \theta$ vertically, Fig. 64.

If air resistance is neglected the only force acting on the body is its weight, and, therefore, the body will be subjected throughout its path to an acceleration g downwards.

Since there is no force acting in the horizontal direction there is no acceleration in this direction and the horizontal velocity remains constant.

Maximum Height

The vertical component v of the velocity at any point P, distance y

above the point of projection, can be obtained from the equation $v^2 = u^2 + 2fs$.

Therefore $\quad\quad\quad v^2 = (u \sin \theta)^2 - 2gy$. Since $f = -g$.

When the body reaches its maximum height—$y = h$—this vertical component velocity will be reduced to zero.

i.e. $\quad\quad\quad 0 = (u \sin \theta)^2 - 2gh$, where h is maximum height,

$$\therefore \; h = \frac{(u \sin \theta)^2}{2g}.$$

Time of Flight

For motion under gravity the vertical distance y travelled in any time t is given by $s = ut + \frac{1}{2}ft^2$.

i.e. $\quad\quad\quad\quad\quad\quad y = (u \sin \theta)t - \frac{1}{2}gt^2$.

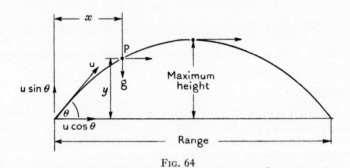

FIG. 64

When the body returns to the ground, the vertical distance which has been travelled is zero. Therefore, when $y = 0$ in this equation, t is equal to the time of flight.

or $\quad\quad\quad\quad\quad 0 = \{(u \sin \theta) - \frac{1}{2}gt\}t,$

i.e. $\quad\quad\quad\quad\quad u \sin \theta = \frac{1}{2}gt$

and $\quad\quad\quad\quad\quad t = \dfrac{2u \sin \theta}{g}.$

(Note also that $y = 0$ at $t = 0$, i.e. at instant of projection.)

Range

The horizontal distance between the point of projection and the point where the body strikes the ground is called the range.

Since the horizontal component of the velocity remains constant throughout the time of flight, the range will equal the horizontal component velocity multiplied by the time of flight.

Then, using $s = vt$, we get :

$$\text{Range} = u \cos \theta \times \frac{2u \sin \theta}{g},$$

or

$$\text{Range} = \frac{2u^2 \sin \theta \cos \theta}{g}.$$

Since $\sin 2\theta = 2 \sin \theta \cos \theta$, we have the alternative form

$$\text{Range} = \frac{u^2 \sin 2\theta}{g}.$$

The actual solution can now be obtained as follows :

(a) Time of Flight. $t = \dfrac{2u \sin \theta}{g} = \dfrac{2 \times 100 \times \sin 30}{g},$

$t = 3\cdot11$ **seconds.**

(b) Range. $R = \dfrac{u^2 \sin 2\theta}{g} = \dfrac{100^2 \times \sin 60}{g},$

$R = 269$ **ft.**

(c) Maximum Height. $h = \dfrac{(u \sin \theta)^2}{2g} = \dfrac{(100 \sin 30)^2}{2g},$

$h = 38\cdot9$ **ft.**

Example 17

If the body in the previous question had been projected not from level ground but from a point on the side of a hill inclined at 30° below the horizontal, what would be its time of flight and the range measured along the side of the hill ?

Solution

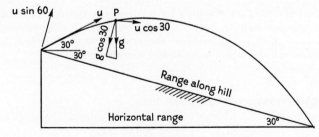

FIG. 65

Time of Flight

As in the previous question motion is still considered *perpendicular to the ground.*

The component of the initial velocity perpendicular to the ground $=u \sin 60°$. Acc. of body at point P perpendicular to the ground $=g \cos 30°$. Applying the equation $s = ut + \frac{1}{2}ft^2$ to motion perpendicular to the ground, s is zero (since the net vertical distance travelled is zero), $u = u \sin 60°$ and $f = -g \cos 30°$.

$$\therefore \ O = u \sin 60 . t - \tfrac{1}{2}gt^2 \cos 30 . t^2.$$

Substituting $u = 100$ ft./sec.,

$$O = 100 \frac{\sqrt{3}}{2}t - \tfrac{1}{2}g\frac{\sqrt{3}}{2}t^2,$$

$$t = 6·25 \text{ seconds.}$$

Range

The horizontal component of the initial velocity, $u \cos 30°$, remains unchanged regardless of the inclination of the ground. Then horizontal distance covered during time of flight is $u \cos 30 \times t$.

$$\therefore \ \text{Horizontal Range} = \left(100 . \frac{\sqrt{3}}{2} \times 6·25\right).$$

But from Fig. 65, Range *along* the hill $= \dfrac{\text{Horizontal Range}}{\cos 30}$,

or $\qquad\qquad \text{Range along hill} = \dfrac{100\dfrac{\sqrt{3}}{2} \times 6·25}{\dfrac{\sqrt{3}}{2}}.$

$$\text{Range along hill} = 625 \text{ ft.}$$

Example 18

Determine the time required for a wheel to acquire a speed of 180 revolutions per minute when starting from rest, with an angular acceleration of 0·3 radians per second per second. How many revolutions will the wheel make before reaching this speed ?

Solution

Angular velocity after 180 revolutions

$$\omega_2 = \frac{2\pi . 180}{60} = 6\pi \text{ rad./sec.}$$

Then $\qquad\qquad\qquad \omega_2 = \omega_1 + \alpha t.$

Initial vel. $\omega_1 = 0$ and $\alpha = 0\cdot 3$.

$$\therefore\ 6\pi = 0 + 0\cdot 3t,$$
$$\therefore\ t = 20\pi$$

$$t = 62\cdot 84 \text{ seconds.}$$

Angular distance θ travelled in this time can be obtained from

$$\omega_2{}^2 = \omega_1{}^2 + 2\alpha\theta,$$
$$\therefore\ 36\pi^2 = 0 + 2 \times 0\cdot 3\theta.$$

$$\theta = 60\pi^2 \text{ radians.}$$

Since there are 2π radians in each revolution,

$$\text{No. of revolutions} = \frac{60\pi^2}{2\pi} = 20\pi,$$

$$\therefore\ \textbf{No. Revolutions} = \textbf{62·84.}$$

15. Derivation of Equations.—The general equations of motion may be derived from the displacement-time curve as follows :

Consider the motion of a body in a straight line (velocity numerically equal to speed) and let the displacement s at any time t be represented by the graph in Fig. 66 (i).

At any given instant the velocity is given by the rate of change of displacement with respect to time which can be obtained from the slope of the displacement-time curve.

$$\text{i.e. } v = \frac{ds}{dt}. \qquad\qquad\qquad\qquad (1)$$

If this slope (velocity) is determined at given time intervals and plotted on a time base the velocity-time curve is obtained as in Fig. 66 (ii).

The slope of this second curve gives the rate of change of velocity with respect to time at any instant—i.e. it gives the acceleration.

$$\therefore\ f = \frac{dv}{dt}. \qquad\qquad\qquad\qquad (2)$$

Combining equations (1) and (2) and eliminating dt, we get the third relationship,

$$v\,dv = f\,ds. \qquad\qquad\qquad\qquad (3)$$

Also from (1) $\qquad v = \dfrac{ds}{dt}$ and differentiating both sides

with respect to t gives :

$$\frac{dv}{dt} = \frac{d^2s}{dt^2}. \qquad \qquad \qquad (4)$$

(When s is differentiated once to give $\dfrac{ds}{dt}$, and this result is again differentiated with respect to time, the result $\dfrac{d^2s}{dt^2}$ is called the second differential coefficient.)

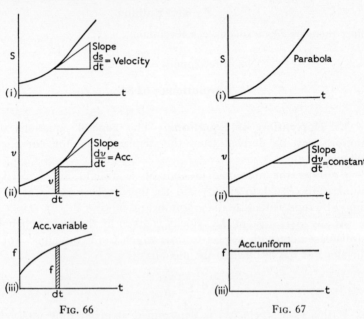

FIG. 66   FIG. 67

Combining equations (2) and (4) we get

$$f = \frac{d^2s}{dt^2}. \qquad \qquad \qquad (5)$$

We, therefore, have the following notation :

s = displacement.

$\dfrac{ds}{dt}$ = rate of change of displacement with respect to time = velocity.

$\dfrac{d^2s}{dt^2}$ = rate of change of velocity with respect to time = acceleration.

Now the area of any element of the velocity-time graph (Fig. 66 (ii)) is given by vdt.

Hence from equation (1) $ds = vdt$.

Integrating both sides of this equation, we get

$$\int ds = \int vdt. \qquad . \qquad . \qquad . \quad (6)$$

But $\int ds =$ displacement and $\int vdt =$ Total area under the velocity-time graph. Therefore :

Displacement = Area under Velocity-Time graph.

In a similar way the area of any element of the acceleration-time graph is given by fdt, and from equation (2) $dv = fdt$.

By integration $\qquad \int dv = \int fdt. \qquad . \qquad . \quad (7)$

i.e. Velocity = Area under Acceleration-Time graph.

Having established the general equations (6) and (7) connecting time with the three variables—displacement, velocity and acceleration—we can now apply them to the particular case of uniform acceleration depicted by the curves in Fig. 67.

For uniform acceleration f is constant as shown at (iii), and for this case equation 7 may be written :

$$\int dv = f \int dt. \qquad . \qquad . \quad (8)$$

Let it be assumed that the displacement $s = 0$ when time $t = 0$.

Also let the initial velocity $= u$ at time $t = 0$.

If v is the velocity at time t, equation (8) becomes

$$\int_{u}^{v} dv = f \int_{0}^{t} dt,$$

$$\therefore v - u = ft,$$

or $\qquad\qquad\qquad v = u + ft. \qquad . \qquad . \quad (9)$

This is the equation of a straight line ($y = mx + c$), see Fig. 67 (ii).

Substituting equation (9) in equation (6)—viz. $\int ds = \int vdt$—we get

$$\int ds = \int (u + ft)dt,$$

and using the given conditions $s = 0$ when $t = 0$,

$$\int_0^s ds = \int_0^t (u + ft)dt,$$

we get $s = ut + \frac{1}{2} ft.^2$

Hence the displacement-time curve for uniform acceleration is a parabola ($s = at + bt^2$)—Fig. 67 (i).

In order to obtain an equation independent of the time we can use the relationship of equation (3), $vdv = fds$.

By integration—using the conditions, velocity $= u$ when $s = 0$ and velocity $= v$ when displacement $= s$.

$$\int_u^v vdv = f \int_0^S ds,$$

or $v^2 = u^2 + 2fs.$

The equations for uniformly accelerated angular motion may be deduced in a similar manner, using the first and second differential coefficients of θ for velocity and acceleration, viz. displacement $= \theta$, angular velocity $\omega = \dfrac{d\theta}{dt}$ and angular acceleration $\alpha = \dfrac{d\omega}{dt}$ or $\dfrac{d^2\theta}{dt^2}$.

Example 19

(a) In a velocity-time curve state what the slope of the curve at any point represents. What is indicated by the area between the velocity-time curve, the time axis, and the ordinates at the two given times ?

(b) A train moving with uniform acceleration passes posts $\frac{1}{2}$ mile apart as follows :

First post, 0 sec. ; second post, 70 sec. ; third post, 120 sec.

Determine the acceleration of the train and its speed when passing the first post.   (N.C.T.E.C.)

Solution

(a) See also article (15), page 63.

(b) Let $u =$ the velocity at the first post (i.e. when $t = 0$). Fig. 69.
    Using equation $s = ut + \frac{1}{2} ft^2$, we have
    for $s = 2640ft$, $2640 = u \cdot 70 + \frac{1}{2} f(70)^2$
    and for $s = 5280ft$, $5280 = u \cdot 120 + \frac{1}{2} f(120)^2$.

Solving these two equations for the two unknowns u and f gives

$$u = 28\cdot9 \text{ ft./sec.} \qquad f = 0\cdot25 \text{ ft./sec.}^2$$

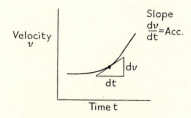

The slope at any point on the velocity-time curve gives the acceleration at that point.

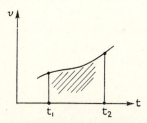

The area under the velocity-time curve during any time interval $t_2 - t_1$ gives the displacement during that interval (See equation 6, article 15.)

FIG. 68

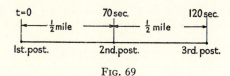

FIG. 69

Example 20

The distance s feet travelled by a body moving in a straight line can be represented by the equation $s = 2t^3 + 1\cdot5t^2 + 8$, where t is the time in seconds. What time will elapse whilst the velocity changes from 9 ft./sec. to 45 ft./sec., and how far will the body travel during this time? What is the acceleration when the body reaches 45 ft./sec.?

Solution

This is a case of variable acceleration (Fig. 66), since the given displacement-time equation is *not* of parabolic form.

$$s = 2t^3 + 1\cdot5t^2 + 8. \qquad . \qquad . \qquad . \qquad (1)$$

Differentiating with respect to time for velocity, we get,

$$\frac{ds}{dt} = 6t^2 + 3t. \qquad . \qquad . \qquad . \qquad (2)$$

To determine the time taken to reach a velocity of 9 ft./sec. we have,

$$\frac{ds}{dt} = 6t^2 + 3t = 9.$$

Then solving the equation $6t^2 + 3t = 9$ or $6t^2 + 3t - 9 = 0$, we get

$$(t-1)(6t+9) = 0.$$

$$t = 1 \text{ second.}$$

Similarly the time to reach 45 ft./sec. from

$$6t^2 + 3t - 45 = 0.$$
$$\therefore (t - 2 \cdot 5)(6t + 18) = 0.$$

$$t = 2 \cdot 5 \text{ seconds.}$$

Hence, the time taken for the velocity to change from 9 ft./sec. to 45 ft./sec. is **1·5 seconds.**

For the distance covered in 1 second substitute $t = 1$ in equation (1).

Thus $\qquad\qquad$ S $= 2 \times 1 + 1 \cdot 5 \times 1 + 8 = 11 \cdot 5$ ft.

Similarly distance covered in 2·5 seconds,

$$\text{S} = 2(2\tfrac{1}{2})^3 + 1 \cdot 5(2 \cdot 5)^2 + 8 = 48 \cdot 6 \text{ ft.}$$

Therefore distance covered in 1·5 seconds $= 48 \cdot 6 - 11 \cdot 5 = \mathbf{37 \cdot 1 \text{ ft.}}$

Differentiating equation (2) with respect to time gives acceleration,

$$\therefore \text{ Acc. } \frac{d^2s}{dt^2} = 12t + 3.$$

When $\qquad\qquad t = 2 \cdot 5$, i.e. when $\dfrac{ds}{dt} = 45$ ft./sec.

$$\frac{d^2s}{dt^2} \text{ is given by } \frac{d^2s}{dt^2} = 12 \times 2 \cdot 5 + 3$$

$$\textbf{Acc.} = \mathbf{33 \text{ ft./sec.}^2}$$

Example 21

A body moving in a straight path with a uniform acceleration of 4 ft./sec.^{2} passes a fixed point A with a velocity of 10 ft./sec. How far is the body from A 6 seconds later, and what is then its velocity?

Solution

For uniform acceleration the displacement-time equation is parabolic —see Fig. 67,

$$\therefore \text{ Acc. } \frac{d^2s}{dt^2} = 4.$$

By integration, $\qquad\qquad \dfrac{ds}{dt} = 4t + C.$ $\qquad . \qquad . \qquad .$ (1)

$\qquad\qquad\qquad\qquad$ where C is a constant of integration.

$\therefore$ Equation (1) gives the velocity at any time t.

If t is taken as zero when the body is at A, then $\dfrac{ds}{dt} = 10$ ft./sec. when $t = 0$.

Substituting these values in equation (1),

$$10 = 0 + C \quad \therefore \quad C = 10 \text{ ft./sec.}$$

Equation (1) becomes $\qquad \dfrac{ds}{dt} = 4t + 10.$.   .   .   .   (2)

Integrating again gives $s = 2t^2 + 10t + D$, where D is a constant of integration.

At time $t = 0$ body is at A, i.e. $s = 0$ at $t = 0$, $\therefore$ D $= 0$.

Then $\qquad\qquad\qquad s = 2t^2 + 10t$ (Parabolic form).

When $t = 6$, $\qquad\qquad s = 2 \times 6^2 + 10 \times 6$

$$= \mathbf{132 \text{ ft.}}$$

Velocity when $t = 6$ from (2)

$$\frac{ds}{dt} = 4 \times 6 + 10$$

$$= \mathbf{34 \text{ ft./sec.}}$$

Distance from A = 132 ft.   **Velocity = 34 ft./sec.**

16. Connection between Linear and Angular Motion.

The distance s travelled by any point moving in a circular path of radius r about a fixed point O—Fig. 70—is given by :

$$s = r\theta.$$

Differentiating the equation with respect to time :

$$\frac{ds}{dt} = r\frac{d\theta}{dt},$$

i.e. Linear velocity $= r \times$ Angular velocity,

or $\qquad\qquad \mathbf{v} = \omega \mathbf{r}.$

Differentiating again,

$$\frac{d^2s}{dt^2} = r\frac{d^2\theta}{dt^2},$$

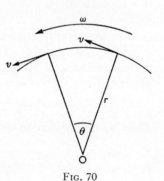

FIG. 70

i.e. Linear Acceleration = r × Angular Acceleration,

or $$f = \alpha r.$$

17. Centripetal Acceleration.—Acceleration is a vector quantity.

If a point moves at uniform speed in a circular path its direction is constantly changing and, therefore, whilst its speed is constant its velocity is changing, i.e. the point has acceleration.

Consider a point A moving in a circular path of radius r with uniform speed v, Fig. 71 (i). After a time dt the point moves to B through the small angle $d\theta$.

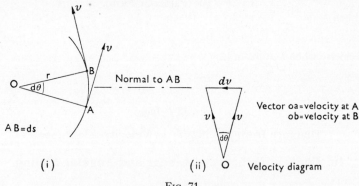

(i)

(ii)   Velocity diagram

FIG. 71

The velocity diagram drawn at (ii) shows that there has been a velocity change dv. This change is in the direction of the normal to the arc AB, and for a very small angle $d\theta$ it may be assumed from the velocity diagram that :

$$vd\theta = dv. \quad \text{(Arc = Angle × Radius.)}$$

Then by differentiation with respect to time,

$$v\frac{d\theta}{dt} = \frac{dv}{dt}. \qquad . \qquad . \qquad . \qquad (1)$$

Also   arc $AB = ds = rd\theta$,

and by differentiation $$\frac{ds}{dt} = r\frac{d\theta}{dt}$$

or $$v = r\frac{d\theta}{dt}. \qquad . \qquad . \qquad (2)$$

From (2)
$$\frac{d\theta}{dt} = \frac{v}{r},$$

which, when substituted in (1), gives

$$\frac{v^2}{r} = \frac{dv}{dt}.$$

But $\frac{dv}{dt}$ is the rate of change of velocity normal to the circular path. Therefore, in the limit, as the angle $d\theta$ tends towards zero, $\frac{v^2}{r}$ will represent the instantaneous acceleration of the point A along the radius OA towards O. This radial acceleration is called *centripetal acceleration*. For uniform speed the centripetal acceleration is constant and always acts towards the centre of curvature O.

Since $v = \omega r$, centripetal acceleration $= \dfrac{v^2}{r} = \omega^2 r$.

It must be realised, particularly for later studies, that this radial or centripetal acceleration is in addition to any angular acceleration that may be present. To illustrate this, consider that the point A in Fig. 72 rotates about O with a variable speed. If

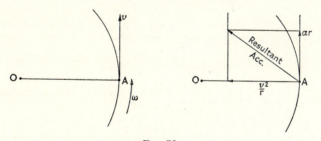

Fig. 72

at the particular instant shown the angular acceleration is α, then the point A will have two components of acceleration.

(i) the tangential acceleration at right-angles to OA $= \alpha r$,

(ii) the centripetal acceleration along the radius OA $= \dfrac{v^2}{r}$, where

v is the particular speed at the instant considered.

Hence, in this case, the instantaneous acceleration of the point

A will be the resultant of the two separate components as shown in Fig. 72.

Example 22

The peripheral speed of a wheel 12 inches diameter increases uniformly from 20 ft./sec. to 40 ft./sec. during 20 revolutions of the wheel. What is the angular acceleration ? If the wheel continues to gain in speed at the same rate, what will be the centripetal acceleration of a point on the periphery when the wheel has turned through a further 10 revolutions ?

Solution

When the linear speed = 20 ft./sec., angular speed $\omega_1 = \dfrac{u}{r} = \dfrac{20}{6/12}$ = 40 rad./sec.

When the linear speed = 40 ft./sec., angular speed $\omega_2 = 80$ rad./sec.

This speed change takes place in 20 revolutions,

$$\therefore \ \theta = 2\pi \ . \ 20 = 40\pi \ \text{radians.}$$

Using $\qquad \omega_2{}^2 = \omega_1{}^2 + 2\alpha\theta,$

$$80^2 = 40^2 + 2\alpha \ . \ 40\pi \quad \therefore \ \alpha = \frac{60}{\pi} \ \text{rad./sec.}^2,$$

i.e. $\qquad\qquad \alpha = \textbf{19·1 rad./sec.}^2$

After a further 10 revolutions with the same angular acceleration, $\omega_3{}^2 = \omega_2{}^2 + 2\alpha\theta$, where ω_3 is the angular speed after 30 revolutions.

$$\omega_3{}^2 = 80^2 + 2 \ . \ \frac{60}{\pi} \ . \ 20\pi$$

θ now equals $2\pi \ . \ 10 \ . \ = 20\pi$ radians,

$$\therefore \ \omega_3{}^2 = 8800.$$

Since Centripetal Acceleration $= \omega_3{}^2 r \ . \ = 8800 \times \dfrac{6}{12}.$

Centripetal Acceleration = 4400 ft./sec.^{2}

18. Relative Velocity.—It should be appreciated that the velocity of a body is only fully defined if it is stated in relation to some other body. This relationship is often implied rather than stated, as, for example, when we speak of a car moving east at 30 miles per hour. Here we imply that the car is moving with the given velocity *relative to the earth*, which, for convenience, we

regard as a fixed body. The fact that the earth is actually rotating
has no effect on us since it is rotating at *constant* velocity—this,
of course, would not be the case if the earth began to accelerate.

Not all velocities are stated relative to the earth as a fixed
body. In many instances the velocity of a body is given relative
to some other body which is itself moving relative to the earth,
and in order to distinguish between these two cases velocities
relative to the earth are called *absolute* velocities and velocities
given in relation to other *moving* bodies are called *relative* velocities. The term *relative velocity* is used, therefore, only when
describing the velocity with which a body appears to be moving
when viewed from another *moving* body.

The simplest problems in relative velocity occur when bodies
are moving in the same straight line. For example, if a car A is
moving due west at 40 m.p.h. and a car B due east at 50 m.p.h.—
Fig. 73 (a)—then the velocity of A relative to B—i.e. the velocity
with which A appears to be moving when viewed from B—will be
90 m.p.h., due west. Conversely, the velocity of B relative to A
will be 90 m.p.h., due east.

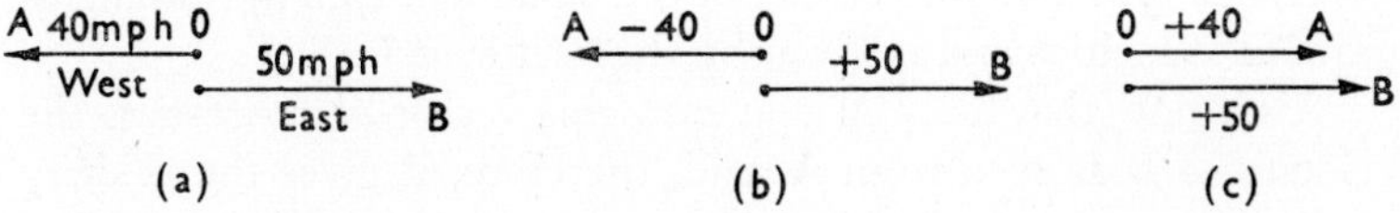

FIG. 73

If the velocity of A is considered negative as at (*b*), then for
the same velocities :

$$\text{Relative velocity} = 50 - (-40) = 90 \text{ m.p.h.}$$

If now the cars both move due east (*c*), then B will overtake A
at 10 m.p.h., and A will appear to be moving backwards at 10
m.p.h., when viewed from B.

$$\text{i.e. Relative velocity} = 50 - 40 = 10 \text{ m.p.h.}$$

These two examples indicate that the relative velocity of
bodies moving in the same straight line is obtained by taking the
algebraic difference of the two absolute velocities.

When the bodies are not moving in the same straight line we
have the case shown in Fig. 74 (i). This diagram illustrates two

A.M.—F

bodies, A and B, which are moving in the same plane with the respective velocities v_a and v_b.

Suppose that the body A is given an extra velocity equal to $-v_a$, and that, in order to retain the same relative velocity of the two bodies, B is treated in the same manner (ii).

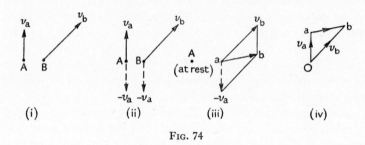

FIG. 74

A's motion is now compounded of the two velocities v_a and $-v_a$, or, in other words, A has been brought to rest.

B will now have a velocity made up of its own velocity v_b and the superimposed velocity $-v_a$.

Hence the resultant velocity of B is obtained from the parallelogram of velocities and is given by the vector ab (iii).

Also, now that body A is at rest, this vector ab represents the velocity of B as seen from A, and, therefore, it gives the velocity of B relative to A. Since this vector is obtained by adding vectorially the velocity of A reversed to the absolute velocity of B, we have the equation :

Velocity of B relative to A = Abs. Velocity of B + (- Abs. velocity of A), or

Velocity of B relative to A = Abs. velocity of B - Abs. velocity of A.

Thus for bodies which are not moving in the same straight line their relative velocity is found by taking their *vector* difference.

If, instead of drawing the parallelogram of velocities (iii), the absolute velocities v_a and v_b are drawn from some pole or earth point O, we get the vector or velocity diagram shown at (iv). Clearly this diagram gives the same vector ab without the necessity of reversing the absolute velocity of A. In this diagram the vectors oa and ob represent the absolute velocities of A and B, respectively.

From this vector diagram we have :

	ab	=	ob	–	oa
	(Velocity B relative to A)		(Abs. Vel. B)		(Abs. Vel. A)
Alternatively	oa	=	ob	–	ab
But vector	ab	=	– vector ba		
∴	oa	=	ob	+	ba
	(Abs. Vel. A)		(Abs. Vel. B)		(Vel. A relative to B)
Similarly	ob	=	oa	+	ab
	(Abs. Vel. B)		(Abs. Vel. A)		(Vel. B relative to A)

Note that for the vector notation used here the first letter of the vector *ab* represents the body which is imagined to be stationary, i.e. for B relative to A (Stationary), letter '*a*' is written first, and for vector *ba* – A relative to B, *b* is written first.

Example 23

A ship A, steaming due east at 20 miles/h., sights a ship B, 10 miles due south. The ship B is steaming north-east at 15 miles/h. Determine the velocity of ship B relative to A. Using this relative velocity, determine (*a*) the nearest distance the ships approach to one another, (*b*) the interval of time after first sighting that elapses before the distance between the ships is a minimum.   (U.L.C.I.)

Solution

To determine the distance between the two ships at any instant, ship A is imagined to be at rest and ship B moving with the relative velocity *ab* as illustrated in the relative space diagram, Fig. 75 (iii). The ships are originally 10 miles apart, and after one hour B will have moved along BD to the point D—a distance of 14·6 miles. The nearest distance that ship B gets to A is clearly given by the perpendicular AC drawn from A onto BD. To the same scale that AB is 10 miles the length AC equals 6·75 miles.

∴ **Nearest distance that ships approach to one another**
= 6·75 miles.

Before the ships are at their minimum distance apart ship B will travel through a distance BC, which, to scale, measures $7\frac{1}{2}$ miles.

If B covers 14·6 miles in one hour relative to A, then the time taken to reach point C is given by $\dfrac{7\frac{1}{2}}{14\cdot6} \times 60 = 30\cdot8$ minutes.

∴ **The time that elapses before the distance between ships**
is a minimum = 30·8 mins.

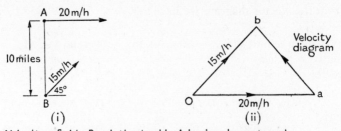

Velocity of ship B relative to ship A is given by vector a b
∴ Relative velocity B to A = 14·6 mile/h. in a direction 41·5° West of North

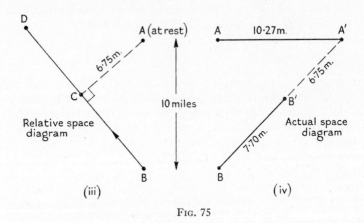

Fig. 75

As both ships are really moving at the same time, the actual space diagram can now be drawn as shown at (iv):

at 20 m./h. A will travel 10·3 miles to A′ in 30·8 minutes,
at 15 m./h. B will travel 7·7 miles to B′ in 30·8 minutes.

Then, as shown, distance A′B′ =6·75 miles.

19. Relative Velocity of Points on a Rigid Link.—So far we have only considered the motion of bodies which move independently of each other. We must examine now the relative velocities of points on a *single* rigid body. As a rigid body is one that can neither bend nor alter its length, then any two given points on it must remain a fixed distance apart. Hence, for the two points A and B at the ends of the link shown in Fig. 76 (i), there can be no relative motion *along* the link, and any velocity of one end relative to the other must be perpendicular to the link. This is illustrated in Fig. 76 (ii) where velocity of A relative to B has been obtained using the methods of the previous article.

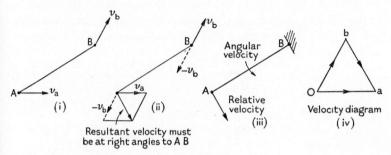

Resultant velocity must
be at right angles to A B

Fig. 76

Whatever the velocities of the two points A and B, if B is imagined to be at rest (iii) the resultant velocity of A must be at right-angles to AB. Fig. 76 (iii) illustrates that, as far as B is concerned, A appears to rotate about B with a tangential velocity given by the vector *ba*.

Now Angular Velocity = $\dfrac{\text{Tangential Velocity}}{\text{Radius}}$ $\left(\omega = \dfrac{v}{r}\right)$.

Therefore, for a link of length AB moving with a tangential velocity *ba*,

$$\text{Angular Velocity} = \frac{ba}{AB}.$$

If the length AB is in feet and the tangential velocity in ft./sec., the angular velocity will be in radians/sec.

Clearly, for the velocity of B relative to A, B would appear to rotate about A (considered fixed) in the same direction with the tangential velocity given by vector *ab*.

It should be appreciated that the link AB can have only one value of angular velocity, although the tangential velocities for different points on the link will vary linearly with their distance from the ends.

Consider the point C on the link in Fig. 77.

The velocity diagram is shown at (ii).

By the argument given above the velocity of C relative to B must be at right-angles to the link.

Hence the velocity of C relative to B—i.e. vector bc—can be found from the relationship $\dfrac{\text{BC}}{\text{BA}} = \dfrac{bc}{ba}$, and the point c located on the vector ba accordingly.

If the absolute velocity of the point C is required, then since o in the velocity diagram is the fixed pole or earth point, we have only to join the point c to o and the absolute velocity of C is given by the vector oc.

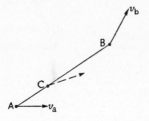

FIG. 77

Example 24

A reciprocating engine has a crank 5 inches long and a connecting rod 14 inches long. The crankshaft rotates at 1400 rev./min. Determine, for the instant when the crank has turned through 50° past the inner dead centre position :

(*a*) the velocity of the piston,
(*b*) the angular velocity of the connecting rod,
(*c*) the magnitude and direction of the velocity of the point on the centre line of the connecting rod 8 inches from the crank pin.

Solution

The reciprocating engine mechanism is drawn to scale in Fig. 78 (i). The crank OB rotates uniformly about O and causes the piston to reciprocate along the fixed cylinder axis.

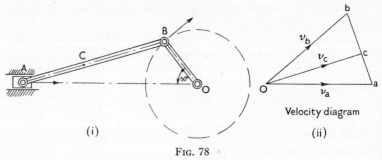

(i)   Velocity diagram   (ii)

FIG. 78

(a) The velocity of point B, which represents the crank pin as well as the end of the connecting rod, can easily be found from the given crank speed. Since $v = \omega r$, Velocity of B $= 2\pi \dfrac{1400}{60} \cdot \dfrac{5}{12}$, $\therefore v_b = 61$ ft./sec.

This velocity of B relative to O will be at right-angles to OB as shown.

The direction of the piston velocity is known but not its magnitude.

Similarly the velocity of A relative to B must be perpendicular to AB, but its magnitude is also unknown.

To draw the velocity diagram : from the pole o in (ii) draw the vector ob to scale ; then a line through b at right-angles to AB will represent the direction of the velocity of A relative to B, and finally a line through o in the direction of the piston velocity will complete the diagram.

From the velocity diagram,

Velocity of piston, oa $= 57$ ft./sec.

Velocity of A Relative to B, $ba = 42$ ft./sec.

(b) The angular velocity of the connecting rod is due to the rotation of A about B (or B about A).

Hence   Angular velocity $= \dfrac{\text{Relative velocity } ba}{\text{Length of connecting rod}} = \dfrac{42}{14/12}$.

Angular velocity of connecting rod $= 36$ rad./sec. (anticlockwise)

(c) If C is the point on the connecting rod 8 inches from B, then point c can be located on the vector ba from the relationship $\dfrac{BC}{BA} = \dfrac{bc}{ba}$.

Then the velocity of point C is represented on the velocity diagram by the vector oc in magnitude and direction.

$\therefore$ **Velocity of point on connecting rod 8 inches from B $= 55.2$ ft./sec.**

Example 25

In the four-bar chain illustrated in Fig. 79 link AD is fixed and the link AB is rotating clockwise at 65 rev./min. For the position shown find the linear velocity of point C and the angular velocities of the links BC and CD.

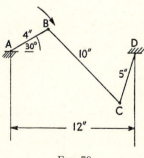

Fig. 79

Solution

In this problem there are two fixed points, A and D, and these will coincide with the pole or earth point o.

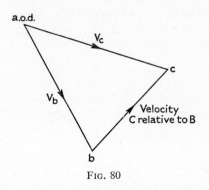

a.o.d.

V_c

c

V_b

Velocity
C relative to B

b

FIG. 80

The velocity of B can be calculated and its direction will be at right-angles to AB,

$$\therefore \text{ Velocity of B} = 2\pi \cdot \frac{65}{60} \cdot \frac{4}{12}$$
$$= 2 \cdot 27 \text{ ft./sec.}$$

To draw the velocity diagram: from the pole o draw vector ob to represent the velocity of B— Fig. 80.

The velocity of C relative to D is known to be at right-angles to CD. Hence through o draw a line at right-angles to CD.

The velocity of C relative to B (imagined fixed) is at right-angles to BC. Therefore through b draw bc at right-angles to BC thus completing the diagram.

Then from velocity diagram :

Velocity of C $= oc = 2 \cdot 47$ **ft./sec.**   **Relative velocity of BC** $= cb$
$= 184$ ft./sec.

The angular velocity of link CD $= \dfrac{\text{Tangential velocity of C}}{\text{Length of link CD}} = \dfrac{2 \cdot 47}{5/12}$.

Angular velocity of CD $= 5 \cdot 93$ **rad./sec.** (anticlockwise)

The angular velocity of link BC $= \dfrac{\text{Tangential velocity C relative to B}}{\text{Length of link BC}}$

$$= \frac{1 \cdot 84}{10/12}.$$

Angular velocity of BC $= 2 \cdot 21$ **rad./sec.** (anticlockwise)

Example 26

Determine the blade inlet angle for an impulse turbine running at a peripheral speed of 450 ft./sec., if the fluid is to enter the blades without shock. The fluid leaves the nozzle with a velocity of 850 ft./sec. at an angle of 20° to the wheel periphery.

Solution

There are three velocities to be considered :

      (*a*) the absolute velocity of the fluid,
      (*b*) the peripheral velocity of the wheel,
      (*c*) the velocity of the fluid relative to the blade.

Then if vector *oa* represents the fluid velocity and vector *ob* the

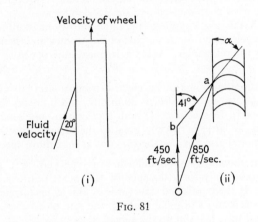

Fig. 81

wheel velocity, we have by the vector equation $oa = ob + ba$, where *ba* represents the velocity of the fluid relative to the blade.

The vector diagram may be drawn from this equation as shown at (ii) in Fig. 81.

If the fluid is to enter the blade smoothly without shock, the velocity diagram indicates that the inlet edge of the blade must be made tangential to the relative velocity *ba*.

Hence, blade angle $\alpha = 41°$ for no shock at entry.

EXERCISES III

   1. A train passes three consecutive mile posts A, B and C. The time taken to travel from A to B is 80 seconds and from B to C 56 seconds. Assuming that the acceleration is uniform find the speeds at A and C in miles/h. and the acceleration in ft./sec.^{2}

   2. A point starts from rest with a constant acceleration which ceases after an interval. It then moves uniformly at 15 ft./sec. for 10 seconds, after which it is uniformly retarded and is brought to rest.

If the whole motion occupies 16 seconds, prove that the distance traversed is 195 feet. The initial acceleration being 5 ft./sec.^{2}, find the final retardation.

3. A train takes 10 minutes to travel from station A to station B, 6 miles away. It receives a uniform acceleration during the first 30 seconds and a uniform retardation during the last 50 seconds of its motion. For the remaining time the train runs at uniform speed. Draw the speed-time graph and determine the uniform speed, the distance travelled at this speed and the acceleration.

A second train also completes the same journey in 10 minutes, being uniformly accelerated from A for part of the journey and then uniformly retarded for the remainder. What is its maximum speed?

N.C.T.E.C.

4. The stopping points of an electric tramcar are 440 yards apart. The maximum speed of the car is 20 m.p.h., and it covers the distance between stops in 75 seconds. If both acceleration and retardation are uniform and the latter is twice as great as the former, find the value of each of them and also how far the car runs at maximum speed.

5. A train travelling at 60 miles/h. was checked by a signal. It decelerated uniformly for a distance of 160 yards until its speed was 15 miles/h. It then travelled 250 yards at this constant speed before accelerating back to its original speed at the rate of 2 ft./sec.^{2} Determine the total time lost due to the signal.

6. A shell is to be fired so that the highest point of its flight is half a mile and its range is 10 miles. Determine the value of the velocity of projection and the inclination of the gun to the horizontal.

7. A shell is fired from a gun in a horizontal direction with a velocity of 1200 ft./sec. The gun is on the side of a hill which has an angle of inclination of 35° to the horizontal. Determine the range of the shell along the side of the hill and also its velocity in magnitude and direction at impact.

8. The position of a particle moving along a horizontal line is given by the equation $s = 10t - 2t^2$ where s is in feet and t is in seconds. Positive direction is to the right.

Determine (a) the velocity of the particle when $t = 1$ second and when $t = 6$ seconds, (b) the displacement of the particle for the time $t = 1$ second to time $t = 6$ seconds, (c) the total distance travelled from time $t = 1$ second to time $t = 6$ seconds.

9. The position of a particle moving along a horizontal line is given by the equation $s = 4t^3 - 24t^2 + 36t + 4$, where s is the distance from the origin in feet and t is the time in seconds. When $t = 0$ the particle is 4 feet to the right of the origin. Determine (a) the times when the velocity is zero, (b) the acceleration when $t = 1 \cdot 5$ seconds.

10. A wheel 10 inches diameter has its peripheral speed increased uniformly from rest to 30 ft./sec. during 40 revolutions of the wheel. What is the angular acceleration ? If the wheel continues to rotate with the same uniform angular acceleration, what will be the centripetal acceleration of a point on the periphery when the wheel has made a further 20 revolutions ?

11. A wheel rotating at 30 rev./min. is uniformly accelerated for 1·5 minutes during which time it makes 75 revolutions. What is the angular velocity of the wheel at the end of this interval and the further interval required for the wheel to reach an angular velocity of 100 rev./min. ?

12. (a) In starting up an electric motor, its speed changes from 500 rev./min. to 2800 rev./min. in 5 seconds. Find the mean angular acceleration in rad./sec.²

(b) If, after switching off, the deceleration is 25 rad./sec.², calculate the time taken to come to rest and the number of revolutions made by the armature. Prove any formulae used. I.Prod.E.

13. A flywheel 4 feet diameter is accelerated from rest at the rate of 0·4 radians/sec.² to a speed of 200 rev./min. It runs for 45 seconds at this speed and is then brought to rest at a uniform rate in 60 revolutions.
Determine :

(a) the retardation in rad./sec.² ;
(b) the total time taken ;
(c) the total number of revolutions made ;
(d) the linear retardation of a point on the rim in ft./sec.

14. A ball is thrown vertically upwards with a speed of 80 ft./sec. and is followed two seconds later by a second ball thrown up in the same way with a speed of 64 ft./sec. Taking 'g' as 32 ft./sec.², determine the relative velocity of the balls immediately after the second ball is thrown and also the distance from the ground where the two balls meet.

15. A ship A, steaming 20° north of east at 18 miles/h., sights another ship B 5 miles due south. The ship B is steaming north-east at 14 miles/h. Determine (a) the velocity of ship B relative to A, (b) the nearest distance the ships approach to one another.

16. To a passenger on a ship travelling due east at 20 m.p.h. another ship 2 miles due south appears to be travelling north-west at 10 m.p.h. Determine (a) the true velocity of the second ship, (b) the nearest distance the ships approach to one another, (c) the interval of time after first sighting that elapses before the distance between the ships is a minimum.

17. A ship A leaves port and steams NE. at 10 m.p.h., and 6 hours later another ship B leaves the same port and steams NW. at 15 m.p.h.

If wireless communication between the two ships is possible up to 300 miles find how far each ship will be from port when communication ceases.

18. In a slider crank mechanism the crank BC is 1 foot long and the velocity of the crank pin B is 10 ft./sec. in a clockwise direction. The connecting rod BA is 4 feet long and the slider A is in line with C, the crank centre. Determine the velocity of the slider when the crank angle has values of 30° and 90° from the inner dead centre position.

19. The crank of a reciprocating engine mechanism is 6 inches long and the connecting rod is 15 inches long, and the crankshaft rotates at 1200 rev./min. Determine for the instant when the crank has turned through 110° past the inner dead centre position :

(*a*) the velocity of the piston ;
(*b*) the angular velocity of the connecting rod ;
(*c*) the magnitude and direction of the velocity of the point on the centre line of the connecting rod 5 inches from the crank pin.

20. ABCD is a four-bar chain with the link AD fixed. The lengths of the links are : AB, 2·5 inches, BC, 7 inches, CD, 4·5 inches, and DA, 8 inches. The crank AB rotates clockwise at 120 rev./min. Determine, for the two positions when the crank AB is in line with the fixed link AD, (*a*) the velocity of the point C, (*b*) the angular velocity of the links CD and BC.

21. An impulse turbine, running with a peripheral speed of 600 ft./sec., has a blade inlet angle of 35°. If the fluid is to enter the blades without shock what angle must the nozzle make with the wheel periphery for a fluid velocity of 1000 ft./sec. ?

Chapter IV

DYNAMICS II—KINETICS

Kinetics is concerned with the *action* of the forces which cause *motion* as distinct from the kinematics in the previous chapter which dealt with the geometry of the motion by itself. In this chapter we shall investigate only those relationships between force and motion which are governed by Newton's Three Laws of Motion * and his subsequent statement of the Law of Universal Gravitation.

20. Newton's Laws of Motion.—(i) A body remains at rest or continues to move with uniform velocity only so long as the resultant of the forces acting on it is zero. (ii) The acceleration of a body is proportional to the resultant force acting on it and is in the direction of the force. (An alternative statement of this law is given in article 29.) (iii) The action and reaction forces between bodies in contact are equal and opposite to each other.

The first law of motion is really only an extension of the second, since if a body is at rest or moving with constant velocity it will clearly have no acceleration and, therefore, will have no resultant force acting on it.

The second law provides the now familiar kinetic equation $F = mf$ where F is the resultant force required to give a body of mass m an acceleration f. The direction of the acceleration f is the same as that of the force F, and $F = mf$ is a *vector* equation.

The third law—fundamental to our whole study of mechanics —tells us that forces always act in *pairs* of equal and opposite forces. It should be observed that this law is not limited to bodies at rest, but applies equally to contacting bodies *in motion* (see article 29 on Impact).

21. The Kinetic Equation.—From Newton's Second Law we know that when a body is subjected to an unbalanced force it

* Newton's laws of motion and the assumption that mass is invariable provide the basic laws of engineering mechanics.

will have an acceleration in the direction of that force.

Let us suppose that we can apply a single unbalanced force F_1 to a body and then measure its resulting acceleration f_1.

Now let the experiment be repeated on the *same* body but with different forces F_2, F_3, etc. The corresponding accelerations being f_2, f_3 and so on. It will very soon become apparent that the ratio of the force to the acceleration remains constant, however many times the experiment is repeated.

Thus $\dfrac{F_1}{f_1} = \dfrac{F_2}{f_2} = \dfrac{F_3}{f_3} =$ some constant C whose value depends on the units employed for F and f.

This constant C is evidently a measure of some property of the body under test which *does not change*.

If this experiment is extended to other bodies, two very important conclusions can be drawn from the results :

(*a*) each body has its own particular value of C,

(*b*) the force required to produce a given acceleration of each body depends directly on the *magnitude* of C.

Evidently this constant C gives a measure of the body's *inertia* or resistance to change in velocity since the higher the value of C the greater is the force required for a given value of acceleration.

This inertia is a constant quantity.

Now the mass of a body gives a measure of its inertia so that C will be proportional to mass.

Since $\dfrac{F}{f} = C$ and C is proportional to mass, we have $\dfrac{F}{f} \propto M$, or $F = KMf$.

By choosing consistent units for force, mass and acceleration we can put K equal to unity and so obtain the fundamental kinetic equation

$$\mathbf{F} = \mathbf{M}f. \qquad . \qquad . \qquad . \qquad (1)$$

The important point to note here is that *both* sides of this equation represent a force.

F is the resultant force applied *to* the body and the product Mf is the inertia force of the body which opposes the change in motion (see Appendix I). When these two forces are equal the body then accelerates in the direction of the force F.

22. The Law of Universal Gravitation.—Newton was also responsible for stating the law governing the mutual attraction between bodies. This law states that every body in the universe is attracted to every other body by a force which is directly proportional to the product of their masses and inversely proportional to the square of their distance apart.

Hence the law of gravitation may be stated thus,

$$F = \frac{\lambda m_1 m_2}{r^2},$$

where F = mutual force of attraction between the masses m_1 and m_2,

r = distance between the centres of the masses,

λ = a constant which is independent of the nature of the masses considered.

For any two bodies of comparable mass situated on the *earth's surface*, the force of attraction between them is negligible. On the other hand, the force of attraction exerted on them by the earth is very much greater, and it is this case—when one of the masses in the above equation is taken as the earth—that we shall now investigate.

Since the earth exerts a force of attraction on all other bodies it is clear from Newton's Second Law that this force will also cause an acceleration. This acceleration, due to the earth's attractive force, is given the symbol g and, therefore, equation (1) becomes

$$F = Mg. \qquad . \qquad . \qquad . \qquad (2)$$

If air resistance is neglected, 'g' will be the gravitational acceleration for bodies falling in vacuum.

Taking M_o to represent the mass of the earth (assumed concentrated entirely at its centre) and M the mass of a body on the earth's surface, we get

$$F = Mg \text{ (Newton's Second Law)}$$

and

$$F = \frac{\lambda M_o M}{r^2} \text{ (Law of Gravitation). (3)}$$

Combining these two equations gives

$$g = \frac{\lambda M_o}{r^2}. \, . \qquad . \qquad . \qquad . \qquad (4)$$

By experiment we have the following approximate values:

Mass of the earth M_o $= 13\cdot2 \times 10^{24}$ lb.
Mean radius of the earth $r = 20\cdot95 \times 10^6$ ft.

$$\lambda = 10\cdot7 \times 10^{-10}\ \frac{\text{ft.}^4}{\text{lb. sec.}^2}.$$

Substituting these values in equation (3)

$$g = \frac{10\cdot7 \times 10^{-10}\ \dfrac{\text{ft.}^3}{\text{lb. sec.}^2} \times 13\cdot2 = 10^{24}\ \text{lb.}}{(20\cdot95 \times 10^6)^2\ \text{ft.}^2},$$

or $g = 32\cdot2$ ft./sec.^{2}

This calculation indicates that every body which is allowed to fall in vacuum at the earth's surface will have an acceleration equal to $32\cdot2$ ft./sec.^{2}

As the earth is flattened at the poles and bulges at the equator, g will vary slightly with position on the earth's surface. However, an average value of $g = 32\cdot2$ ft./sec.^{2} is accurate enough for most engineering problems.

23. Weight

In the kinetic equation $F = Mf$ the force F may act in any direction we like, providing it is remembered that the resulting acceleration will always be in the same direction as F.

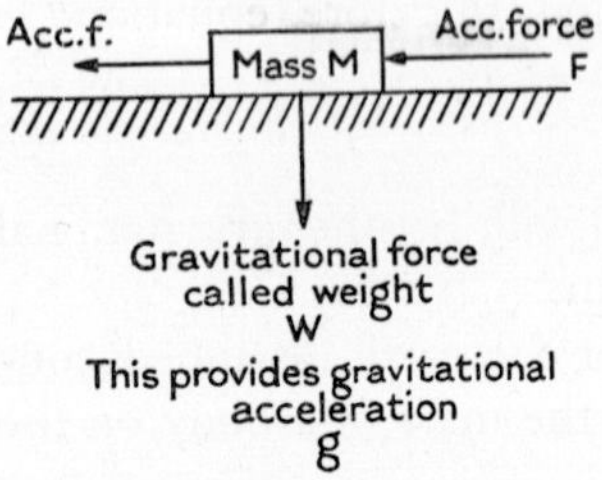

Acc.f.   Mass M   Acc.force
                              F

Gravitational force
called weight
W
This provides gravitational
acceleration
g

Fig. 82

The particular *force* which attracts a body towards the earth's surface is called its WEIGHT, and, therefore, it is the weight of the body which provides its gravitational acceleration.

Then for any body of mass M at the earth's surface it is the weight of the body W which will cause it to have an acceleration f equal to g—Fig. 82.

Since **Weight is a force**, we can substitute W for F in the kinetic equation to give

$$W = Mg. \qquad . \qquad . \qquad . \qquad (5)$$

Now the law of gravitation shows quite clearly that the force of

attraction or weight of any body varies inversely with the square of
its distance from the earth's centre, i.e. $W = \dfrac{\lambda M_o M}{r^2}$.

Evidently then the weight of a body is not a constant quantity.

Further, we know that g is only 32·2 ft./sec.² at the earth's
surface and is also a value which depends on the relative position
of the body from the earth, i.e. $g = \dfrac{\lambda M_o}{r^2}$.

Combining these two relationships we get

$$\frac{W}{g} = M,*$$

which obviously follows from equation (4).

This indicates that whilst both W and g vary with position
relative to the earth their ratio remains constant and equal to the
mass M. Hence W/g is also a measure of inertia and provides
the alternative form of the kinetic equation, viz.

$$F = \frac{W}{g}f.$$

24. Units.—If we take unit force as that force which, when
acting on a body of unit mass, gives to it unit acceleration, the
value of the constant K in the equation $F = Kmf$ is conveniently
made equal to unity.

E.g. if (1 unit of force) = K × (1 unit of mass) × (1 unit of acc.),
then K = unity.

Choosing units in this way means that the units of force, mass
and acceleration *will not be independent of each other*, so that great
care must be exercised when using the kinetic equation.

The two systems of units in common use are the *absolute*
system and the *gravitational* or *engineers'* system.

In the absolute system the acceleration (length/time²) and the
mass are considered as fundamental units and the unit of force is
derived from them.

The mass is measured in lb. mass and the acceleration in
ft./sec.²

Then 1 unit of force = 1 lb. mass $\times \dfrac{1 \text{ ft.}}{\text{sec.}^2}$.

---

* Since g varies with distance from the earth's surface, bodies outside the
earth's gravitational field will have mass but no weight.

A.M.—G

This gives the unit of force as the $\dfrac{\text{ft.lb. mass}}{\text{sec.}^2}$, which is called the POUNDAL.

In the gravitational system the acceleration and the force are considered as fundamental units and the unit of *mass* is derived from them.

The force is measured in lb. force and the acceleration in ft./sec.^{2} Since weight is a force the unit of force is taken as the lb. wt. (pound weight).

Then in this system

$$1 \text{ lb. wt.} = 1 \text{ unit of mass} \times \frac{1 \text{ ft.}}{\text{sec.}^2}.$$

and this gives the unit of mass as the

$$\frac{\text{lb. wt. sec.}^2}{\text{ft.}}, \text{ which is called the SLUG.}$$

We can now apply both these systems of units to the equation

$$W = mg, \text{ in which the weight W is a force.}$$

In the gravitational system we have

1 lb. wt. acts on 1 slug mass to give 1 ft./sec.^{2} acc.

In the absolute system

1 poundal acts on 1 lb. mass to give 1 ft./sec.^{2} acc.,

i.e.   1 poundal = 1 lb. mass $\times$ 1 ft/sec.^{2} acc.

Multiplying both sides of this equation by g, we get

$$g \text{ poundals} = g \text{ lb. mass} \times 1 \text{ ft./sec.}^2$$

or   **g poundals acts on g lb. mass to give 1 ft./sec.^{2} acc.**

Comparing the relationships given by both systems it is evident that we have

$$g \text{ poundals} \equiv 1 \text{ lb. wt.}$$

$$g \text{ lb. mass} \equiv 1 \text{ slug.}$$

25. Linear Translation.—If a body is acted upon by any number of external forces and the resultant of these forces passes through its centre of gravity, the body is said to have a translatory motion—Fig. 83. As the resultant force passes through the centre of gravity the *moment* of the resultant force about every axis through the centre of gravity will be zero for this kind of motion.

F has no moment about CG. There-fore, body does not rotate and has translatory motion only.

All particles of the body have the same acceleration.

FIG. 83

Example 27

A train, weight 250 tons, moving with a uniform velocity of 40 m.p.h. along a horizontal track, begins to climb up an inclined bank having a slope of 1 in 80. During the climb the engine exerts a constant tractive force of 5000 lb. whilst the resistances to motion remain constant at 15 lb. per ton weight of the train. Determine how far the train will climb up the bank before coming to a standstill. Give your answer in feet.   (N.C.T.E.C.)

Solution

The forces opposing the motion of the train along the inclined bank are :

(i) The component of the train's weight down the plane,

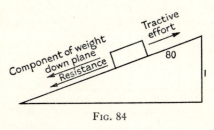

FIG. 84

i.e. $(250 \times 2240)\dfrac{1}{80} = 7000$ lb.

(ii) The frictional resistance at 15 lb./ton—acting down the plane,

i.e. $15 \times 250 = 3750$ lb.

Since the train is retarded and finally brought to rest, the sum of these two forces will exceed the tractive effort causing motion up the bank.

Hence   Retarding Force $= 7000 + 3750 - 5000 = 5750$ lb.

If f is the retardation, we have :

$$\text{Retarding Force} = \text{Mass} \times f$$

or
$$5750 = \frac{250 \times 2240}{32\cdot2} \times f$$

Mass in slugs.   Retarding Force in lb. wt.

and
$$f = 0\cdot33 \text{ ft./sec.}^2$$

Now the train commences the incline with an initial velocity of 40 m.p.h.

or
$$40 \times \frac{44}{30} = 58\cdot7 \text{ ft./sec.}$$

Then, using the equation $v^2 = u^2 - 2fs$:

Where v is the final velocity of the train (which is zero),
$\quad u$ is the initial velocity up the incline,
$\quad s$ the distance travelled before coming to rest,

we have
$$O = (58\cdot7)^2 - 2 \times 0\cdot33\ s.$$

$$\therefore\ s = \mathbf{5230\ ft.}$$

Example 28

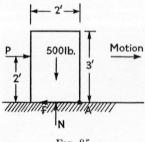

Fig. 85

The block shown in Fig. 85 rests on a horizontal plane and the coefficient of friction between block and plane is 0·3. Determine the force P required :

(a) to cause the block to just slide across the plane ;
(b) to tip the block about the right-hand corner A.

Also determine the maximum acceleration which can be given to the block without it tipping.

Solution

(a) When the block is about to slide the friction force F will reach its limiting value μN.

For vertical equilibrium N = 500 lb.
$\therefore$ Max. friction force F $= \mu N = 0\cdot3 \times 500 = 150$ lb.

Also for horizontal equilibrium P = F = 150 lb., and, therefore, 150 lb. is the force which is just sufficient to give uniform motion to the block.

(*b*) The force P to cause the body to tip about point A can be found by taking moments of the forces about that point. See Fig. 86.

At the instant of tipping N will act through point A.

The clockwise moment about A = Anti-clockwise moment about A,

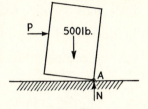

FIG. 86

or
$$P \times 2 = 500 \times 1,$$
$$P = 250 \text{ lb.}$$

Hence the maximum value of P which can be applied without tipping the block is 250 lb., whilst the minimum value of P required to cause sliding is 150 lb.

For values of P within these two figures the block will be accelerated to the right.

Then for accelerated motion we have P – F = Accelerating force.

Now maximum value of P is 250 lb.—i.e. when block is on verge of tipping, and limiting friction conditions exist F = 150 lb.

∴ Maximum accelerating force (P – F) = 250 – 150 = 100 lb.

Then since Force = Mass × Acceleration,

$$100 = \frac{500}{32 \cdot 2} f, \text{ where } f \text{ is the maximum possible acceleration,}$$

∴ $f = 6 \cdot 44$ ft./sec.²

26. Translation in a Curved Path

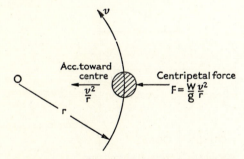

Centripetal Force required to keep body moving in circular path acts in direction of Acceleration.

FIG. 87

When a body moves with uniform velocity in a *curved* path of constant radius of curvature the resultant force acting on the body must impose a radially inward acceleration $\dfrac{v^2}{r}$ on it as shown in Chapter III. Since Force = Mass × Acc. still applies, we have :

$$\text{Centripetal Force} = \text{Mass} \times \text{Centripetal Acceleration,}$$

or
$$F = \frac{W}{g}\frac{v^2}{r}.$$

The following problems illustrate the application of this equation to uniform motion in a circular path.

Example 29

A centrifugal clutch consists of 4 blocks, each weighing 4 lb., rotating in the housing and connected to the axis of rotation by springs of stiffness 60 lb. per inch. When stationary the centre of gravity of each weight is 5 inches from the axis of rotation and there is a 1-inch clearance between the blocks and the housing, which is 13 inches internal diameter. Calculate (*a*) the speed at which the clutch will begin to transmit power, (*b*) the power transmitted at 500 rev./min. if the coefficient of friction between the blocks and the housing is 0·25.

Solution

(*a*) The clutch will just begin to transmit power when each block makes contact with the housing, i.e. when the blocks have moved outwards 1 inch and the radius of rotation of the CG of the blocks is 6 inches from the axis of rotation.

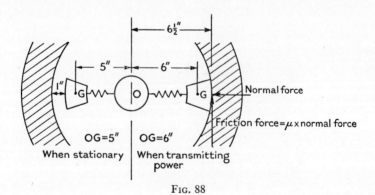

FIG. 88

For this condition the centripetal inward force on each block is supplied by its spring which will have extended 1 inch.

$$\text{Spring force} = \text{Spring stiffness} \times \text{extension}$$

$$= 60 \, \frac{\text{lb.}}{\text{in.}} \times 1 \text{ in.} = 60 \text{ lb.}$$

Then $\qquad\qquad 60 \text{ lb.} = \dfrac{w}{g} \dfrac{v^2}{r} = \dfrac{w}{g} \omega^2 r \text{ . lb.}$

or $\qquad\qquad 60 = \dfrac{4}{32 \cdot 2} \times \omega^2 \times \dfrac{6}{12} \text{ in.}$

$$\therefore \; \omega = \mathbf{30 \cdot 98 \; rad./sec.} = \mathbf{296 \; rev./min.}$$

(b) At 500 rev./min. the centripetal force on each block

$$= \frac{4}{32 \cdot 2} \left(\frac{500 \times 2\pi}{60} \right)^2 \cdot \frac{6}{12},$$

$$= 171 \cdot 25 \text{ lb.}$$

Of this force 60 lb. is supplied by the spring tension and the remainder, i.e. $171 \cdot 25 - 60 = 111 \cdot 25$ lb., will be supplied by the reaction between the block and the housing.

Hence the normal force between each block and housing $= 111 \cdot 25$ lb.

Tangential frictional force $= \mu \times$ Normal Force $= 0 \cdot 25 \times 111 \cdot 25$,

Frictional force $= 27 \cdot 81$ lb.

This frictional force acts at the inner surface of the housing, i.e. at a radius of $6\frac{1}{2}$ inches.

Then frictional torque/block $= 27 \cdot 81 \times \dfrac{6\frac{1}{2}}{12} \text{ in.} = 15 \text{ lb. ft.}$

Total frictional torque for 4 blocks $= 4 \times 15 = 60$ lb. ft.

$$\text{HP} = \frac{2\pi \text{NT}}{33000},$$

where T is the frictional torque in lb. ft. and $N = 500$ rev./min.

$$\therefore \; \text{HP} = \frac{2\pi \times 500 \times 60}{33000}.$$

$$\text{HP} = 5 \cdot 71.$$

The Conical Pendulum

The arrangement which is shown in Fig. 89 (a) constitutes a conical pendulum and consists of a small weight W suspended by a light arm of length l. The small weight or bob is made to revolve in a horizontal circular path of radius r with constant angular velocity ω. For any given speed of rotation, the inclination θ of the arm to the vertical, the height of the bob h, and the

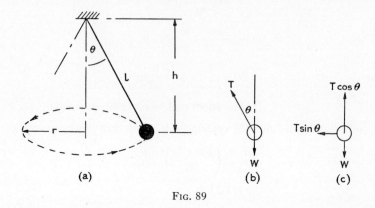

Fig. 89

radius of rotation r will have fixed values. If T is the tension in the arm, then the only forces acting *on* the bob will be this tension and its weight W—Fig. 89 (b).

The tension T can be resolved into two components as shown at (c). These are :

(i) a vertical component T cos θ equal to W ;

(ii) a horizontal component T sin θ which supplies the radial inward force, i.e. the centripetal force necessary to keep the bob moving in a circular path.

We, therefore, have the two relationships :

$$\text{T}\cos\theta = \text{W} \qquad . \qquad . \qquad . \qquad . \qquad \text{(i)}$$

and

$$\text{T}\sin\theta = \frac{\text{W}}{g}\omega^2 r \qquad . \qquad . \qquad . \qquad \text{(ii)}$$

where $\omega^2 r$ is the centripetal acceleration since $v = \omega r$.

Dividing (i) by (ii)

$$\frac{\cos\theta}{\sin\theta} = \frac{1}{\tan\theta} = \frac{g}{\omega^2 r},$$

or

$$\frac{r}{\tan\theta} = \frac{g}{\omega^2}.$$

But from the geometry of Fig. 89 (a), $\dfrac{r}{\tan \theta} = h$,

$$\therefore \ h = \frac{g}{\omega^2}.$$

This indicates that the height h of a conical pendulum from the plane of rotation to the point of suspension depends only on the speed of rotation and is independent of the length of the arm l.

Also from the relationship $T \sin \theta = \dfrac{W}{g} \omega^2 r$,

we have $\qquad\qquad T = \dfrac{W}{g} \omega^2 \cdot \dfrac{r}{\sin \theta}.$

Since $\qquad \dfrac{r}{\sin \theta} = l$ we have tension $T = \dfrac{W}{g} \omega^2 l.$

Example 30

(a) Assuming the formula for centripetal acceleration, show that the height h of a conical pendulum is given by $h = g/\omega^2$ where ω is the angular velocity of the pendulum in radians per sec.

(b) A simple governor has arms 10 inches long pivoted on the axis of the governor spindle. Find the governor speed in r.p.m. when the arms are inclined to the spindle of the governor at an angle of 25°.

(c) What would be the angle of inclination of the arms to the governor spindle when the speed of the governor increases to 70 r.p.m. ?

(N.C.T.E.C.)

Solution

(a) $h = g/\omega^2$ as deduced above.

(b)

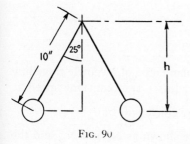

Fig. 90

From the arrangement of the arms shown in Fig. 90 :

$$\cos 25° = \frac{h}{10} \text{ or } h = 10 \cos 25°.$$

Then using the relationship $h = g/\omega^2$ we have

$$10 \cos 25° = \frac{32 \cdot 2 \times 12}{\omega^2},$$

g must be in in./sec.² if h is in inches.

and
$$\omega = \sqrt{\frac{38\cdot64}{\cos 25°}},$$

$$\omega = 6\cdot54 \text{ rad./sec.}$$

But
$$\omega = \frac{2\pi \cdot \text{rev./min.}}{60}.$$

Speed = 62·5 rev./min.

(c) When the speed is 70 rev./min.,

$$\omega = \frac{2\pi \cdot 70}{60} = 7\cdot34 \text{ rad./sec.}$$

Then from (b) we have :

$$\omega^2 = \frac{38\cdot64}{\cos \theta}$$

or
$$\cos \theta = \frac{38\cdot64}{7\cdot34^2} = 0\cdot72.$$

$$\theta = \mathbf{44°.}$$

Vehicle on a Curved Horizontal Track

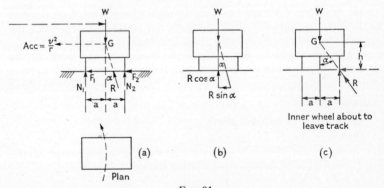

FIG. 91

Fig. 91 (a) shows the end and plan views of a vehicle travelling on a curved horizontal track of radius r at a constant speed v. The centripetal acceleration of the vehicle will be $\frac{v^2}{r}$.

The normal wheel reactions are shown as N_1 and N_2, and the road friction forces as F_1 and F_2.

These four forces may be conveniently replaced by their single resultant R, which will be inclined to the normal to the road surface at some friction angle α.

For no rotation about the CG this resultant R will pass through G, as shown in the diagram, Fig. 91 (a).

Hence the only forces acting on the vehicle may be represented by the weight W and the resultant R.

Clearly, only the *horizontal* frictional component of R can supply the inward centripetal force $\dfrac{W}{g}\dfrac{v^2}{r}$ necessary to keep the vehicle moving in a circular path.

Then from Fig. 91 (b)

$$R \sin \alpha = \frac{W}{g}\frac{v^2}{r}. \qquad \qquad (1)$$

Also for vertical equilibrium we have

$$R \cos \alpha = W. \qquad \qquad (2)$$

Dividing equation (1) by equation (2) gives

$$\tan \alpha = \frac{v^2}{gr}. \qquad \qquad \text{equation (A)}$$

Now the maximum permissible value that the angle α can have is φ—the angle for limiting friction conditions—and when this is reached, *slipping* of the wheels will occur and the centripetal force $\dfrac{W}{g}\dfrac{v^2}{r}$ will have attained its maximum value.

Hence, when

$$\alpha = \phi, \ \tan \phi = \mu = \frac{v^2}{gr},$$

where μ is the coefficient of friction,

$$\therefore \ v^2 = \mu g r$$

and
$$v = \sqrt{\mu g r}. \qquad \qquad \text{equation (B)}$$

This is the maximum velocity for a bend of given radius and skidding will occur if this value is exceeded.

Depending on the design of the vehicle it is possible for it to overturn about its outer wheels long before the speed is sufficient to cause skidding.

If overturning is to take place it is evident that the reaction on the inner wheels N_1, and consequently F_1, must be zero, and for this condition the whole weight of the vehicle will be transferred to the outer wheels. Fig. 91 (c) illustrates this case, where it is shown that the angle α has been increased until R acts entirely at the outside wheels.

From the geometry of Fig. 91 (c) $\tan \alpha = \dfrac{a}{h}$ where 'a' is half the wheel base and h the height of the CG above the track.

Substituting $\tan \alpha = \dfrac{a}{h}$ in equation (A),

we get
$$\frac{a}{h} = \frac{v^2}{gr},$$

and hence
$$v^2 = \frac{gar}{h}$$

and
$$v = \sqrt{\frac{gar}{h}}. \qquad \text{equation (C)}$$

Therefore, if $v > \sqrt{\dfrac{gar}{h}}$ the vehicle will overturn.

Equation (C) indicates that the tendency to overturn is clearly related to the dimensions of the vehicle since v may be increased either by lowering the height of the centre of gravity h or by increasing the width of the wheel base, 2a.

Summary

If $v > \sqrt{\mu gr}$, equation (B) vehicle will skid.

If $v > \sqrt{\dfrac{gar}{h}}$, equation (C) vehicle will overturn.

Example 31

Show that when a body moves in a circular path of radius R with a linear velocity V, the centripetal acceleration is equal to V^2/R.

The width between the wheels of a motor car is 5 feet, and when the car is loaded, its centre of gravity is $2\frac{1}{2}$ feet above the ground in a plane midway between the wheels. If the maximum coefficient of friction between the tyres and the road is 0·5, show that the car will side slip and not overturn when cornering at speed on a level road. Calculate the greatest speed at which a bend of 50 feet mean radius can be negotiated without side slip. (U.E.I.)

Solution

The centripetal acceleration V^2/R is derived in Chapter III, article 17.

For overturning to occur the reaction R must occur wholly at the outer wheels, and from Fig. 92,

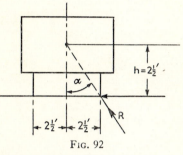

$$\tan \alpha = \frac{a}{h} = \frac{2\frac{1}{2}}{2\frac{1}{2}} = 1 \text{ for the}$$

overturning condition.

Now overturning can only occur if the friction force is great enough to keep the outer wheels from slipping, but since $\tan \phi = \mu = 0.5$ is clearly less than $\tan \alpha$ (i.e. $0.5 < 1$) the friction force is insufficient and the car will, therefore, side slip long before overturning is possible.

Fɪɢ. 92

The greatest speed without side slip is given by equation B :

$$\text{i.e. } V = \sqrt{\mu g R}.$$

Substituting the given values $\mu = 0.5$ and $R = 50$ feet, we get

$$V = \sqrt{0.5 \times 32.2 \times 50} = \textbf{28.35 ft./sec.}$$

$$\text{speed in M.P.H.} = \frac{28.35 \times 3600}{5280} = \textbf{19.33 m.p.h.}$$

Example 32

A motor vehicle of total weight 12,500 lb. starts the ascent of a 1 in 50 gradient at a speed of 20 m.p.h., and maintains a constant tractive effort of 750 lb. throughout the total length of 2000 feet. The road resistance amounts to 75 lb. per ton.

After the ascent it has to travel along a straight level stretch of 500 feet in length before taking a bend of 150 feet radius. The roadway at the curve is unbanked and the coefficient of friction against side slip is 0·4. Determine the uniform tractive effort that is permissible on the 500-foot length if the vehicle is not to reach a speed at the curve in excess of that at which skidding would occur. (I.Mech.E.)

Solution

Consider the motion along the inclined track AB. (Fig. 93)

The frictional resistance at 75 lb./ton

$$= \frac{12500}{2240} \times 75 = 418 \text{ lb.}$$

The component of vehicle's weight acting down incline

$$= 12500 \times \frac{1}{50} = 250 \text{ lb.}$$

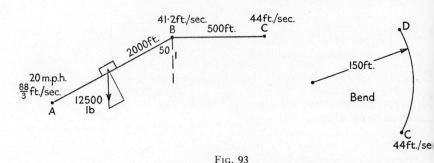

FIG. 93

Then, since Tractive Effort = Resistance to Motion + Accelerating Force, we have $750 = 418 + 250 +$ Accelerating Force :

$$\therefore \text{ the Accelerating Force} = 82 \text{ lb.}$$

This accelerating force will accelerate the vehicle up the incline from A to B, and the value of the acceleration f is given by :

$$\text{Acc. Force} = \text{Mass} \times f$$

or

$$82 = \frac{12500}{32 \cdot 2} \times f,$$

$$\therefore f = 0 \cdot 211 \text{ ft./sec.}^2$$

The velocity of the vehicle at A is 20 m.p.h. or $\dfrac{88}{3}$ ft./sec.

Then, since the distance along the incline s is 2000 ft., we have

$$v_{\mathrm{B}}{}^2 = v_{\mathrm{A}}{}^2 + 2fs$$

or

$$v_{\mathrm{B}}{}^2 = \left(\frac{88}{3}\right)^2 + 2 \times 0 \cdot 211 \times 2000.$$

Hence

$$v_{\mathrm{B}} = 41 \cdot 2 \text{ ft./sec.}$$

This is the velocity with which the vehicle enters the horizontal 500-foot stretch of road, i.e. v_{B}.

Consider the motion along the curved track CD

The maximum velocity which the vehicle can have along the curved track without skidding is given by :

$$v = \sqrt{\mu g r}, \text{ where } r \text{ is the radius of curvature.}$$

Hence maximum velocity at C given by

$$v_C = \sqrt{0.4 \times 32.2 \times 150} = \mathbf{44 \text{ ft./sec.}}$$

Consider the motion along the horizontal track BC

The vehicle enters the horizontal stretch with velocity $v_B = 41.2$ ft./sec. and leaves it with velocity $v_C = 44$ ft./sec.

Hence acceleration along this stretch, for which S $= 500$ feet, is given by $v_C{}^2 = v_B{}^2 + 2f_H s$, where f_H is the horizontal acceleration.

$$\therefore \ 44^2 = 41.1^2 + 2f_H \cdot 500,$$

and $\qquad\qquad f_H = 0.232 \text{ ft./sec.}$

Acc. Force to give this acceleration F $=$ Mass $\times$ Acceleration

$$= \frac{12500}{32.2} \times 0.232 = 90 \text{ lb.}$$

Then tractive effort along horizontal $=$ Acc. Force

$$+ \text{ Horizontal resistance}$$

$$= 90 + 418.$$

Tractive effort permissible $= 508$ lb.

Vehicle on an Inclined Curved Track

We shall now investigate the condition for skidding and over-turning when the track is banked at an angle θ as in Fig. 94.

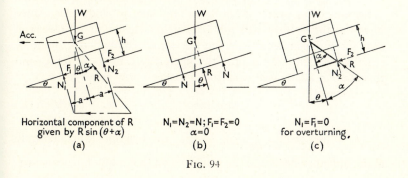

Horizontal component of R given by R sin $(\theta + \alpha)$
(a)

$N_1 = N_2 = N; F_1 = F_2 = 0$
$\alpha = 0$
(b)

$N_1 = F_1 = 0$
for overturning.
(c)

Fig. 94

The resultant reaction R is still inclined at angle α to the normal of the track but now makes an angle of $(\theta + \alpha)$ with the line of action of W—see Fig. 94 (a).

Proceeding as for the horizontal track we get

$$R \sin (\theta + \alpha) = \frac{W}{g} \frac{v^2}{r}, \qquad . \qquad . \qquad (1)$$

and

$$R \cos (\theta + \alpha) = W. \qquad . \qquad . \qquad (2)$$

Dividing equation (1) by equation (2) gives

$$\tan (\theta + \alpha) = \frac{v^2}{gr}.$$

This equation should be compared with equation (A) for a horizontal track.

Rearranging and using the expansion for $\tan (\theta + \alpha)$,

$$v^2 = gr \cdot \frac{\tan \theta + \tan \alpha}{1 - \tan \theta \tan \alpha}. \qquad . \qquad \text{equation (D)}$$

In the case of a banked track there are two conditions which need to be considered—see Fig. 94 (b) and (c).

(a) When $\alpha = 0$, R will be normal to the track, and hence there will be no road friction, i.e. $F_1 = F_2 = 0$ and $N_1 = N_2$.

Then from equation (D), putting $\tan \alpha = 0$, we get

$$v^2 = gr \cdot \tan \theta$$

or

$$\tan \theta = \frac{v^2}{gr}.$$

The angle θ in this equation gives the *angle of banking* for which —friction being zero—there is no tendency to skid. Also since $N_1 = N_2$, the vehicle will not overturn. It should be observed that for this condition it is the horizontal components of the N reactions which *alone* supply the necessary centripetal force, and this state of affairs will exist for one speed only as given by :

$$v = \sqrt{gr \tan \theta}. \qquad . \qquad . \qquad \text{equation (E)}$$

(b) When $\tan \alpha = \frac{a}{h}$—Fig. 94 (c)—the reaction R acts entirely at the outside wheels and the vehicle is on the point of overturning.

Putting $\tan \alpha = \dfrac{a}{h}$ in equation (D) we get

$$v^2 = gr \frac{\tan \theta + a/h}{1 - \tan \theta . \dfrac{a}{h}}.$$

This equation implies that friction is sufficient to allow the vehicle to rotate about the outside wheels :

i.e. $\tan \phi > \tan \alpha$ or $\mu > a/h$.

If this is not the case and friction is insufficient, skidding will occur before the vehicle can overturn and $\tan \phi < \tan \alpha$ or $\mu < a/h$.

Hence the limiting condition for skidding is, therefore, when $\tan \alpha = \tan \phi = \mu$ and the speed for skidding is given by :

$$v^2 = gr . \frac{\tan \theta + \mu}{1 - \mu \tan \theta}. \qquad \text{equation (F)}$$

Example 33

A car travels round a banked bend having a mean radius of 120 feet. If the angle of banking is 30° and the coefficient of friction between the tyres and the ground is 0·7, determine the maximum speed of the car in m.p.h. when side slip is about to occur.

Solution

When side slip is about to occur limiting friction conditions will exist at the road wheels—i.e. $\tan \alpha = \tan \phi = \mu$—and equation (F) will apply.

Hence $\qquad v^2 = gr \dfrac{(\tan \theta + \mu)}{1 - \mu \tan \theta}$, where $\theta = 30°$, and $\mu = 0·7$.

Then $\qquad v^2 = 32·2 \times 120 \dfrac{(0·5774 + 0·7)}{1 - 0·7 \times 0·5774}$,

$v^2 = 8228$ and $v = 90·7$ ft./sec.

Therefore maximum possible speed for no side slip is 90·7 ft./sec. or **61·85 m.p.h.**

Example 34

A bicycle and rider together weigh 200 lb. Find the angle which the machine and rider must make with the horizontal road when

A.M.—H

travelling round a curve of 30-foot radius at 10 m.p.h. without side slip. What is then the frictional force between the tyres and the road ?

Determine also the frictional force for the same speed and radius of curvature when the road is banked 10° in the cyclist's favour.

Solution

Consider motion along the horizontal road. (Fig. 95)

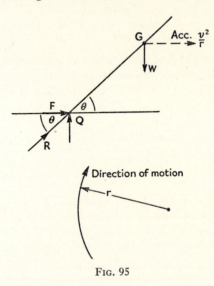

FIG. 95

The horizontal inward centripetal force necessary to keep the bicycle moving in a curved path can only be supplied by the friction force F.

Therefore $\quad F = \dfrac{W}{g}\dfrac{v^2}{r} \quad v = 10 \text{ m.p.h.} = \dfrac{88}{6} \text{ ft./sec.}$

$$r = 30 \text{ ft.}$$

Then $\quad F = \dfrac{200}{32 \cdot 2}\left(\dfrac{88}{6}\right)^2\dfrac{1}{30} = \textbf{44·6 lb.}$

For the bicycle and rider to be in rotational equilibrium the resultant reaction R must pass through G, i.e. $\tan\theta = \dfrac{Q}{F}$.

But for vertical equilibrium $Q = W = 200$ lb.

$$\therefore \quad \tan\theta = \dfrac{200}{44 \cdot 6} \text{ and } \theta = \textbf{77° 25'.}$$

Consider the motion along the banked road. (Fig. 96)

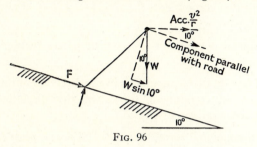

Fig. 96

The horizontal centripetal force $\dfrac{W}{g} \dfrac{v^2}{r}$ can be resolved into two components—one parallel with the road and the other at right-angles to it.

The component parallel to the road $\dfrac{W}{g} \dfrac{v^2}{r} \cos 10°$ will be supplied by the friction force F *assisted* by the component of W parallel to the road,

i.e. $F + W \sin 10° = \dfrac{W}{g} \dfrac{v^2}{r} \cos 10°,$

$$\therefore \ F = 44{\cdot}6 \cos 10 - 200 \sin 10.$$

$$F = 9{\cdot}2 \text{ lb.}$$

27. Rotation about a Fixed Axis.—In the previous article it was shown that when a body has translatory motion only the resultant of the external forces acting on it passes through the *centre of gravity* of the body.

We shall now investigate the case when this resultant force does *not* act through the centre of gravity.

Consider the body in Fig. 97 which is rotating in a horizontal plane about a fixed axis O with angular

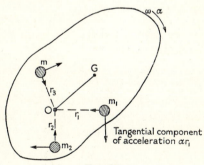

Fig. 97

acceleration α and instantaneous angular velocity ω. (The case of rotation in a vertical plane is dealt with in Chapter V.)

The acceleration of any elemental mass m_1 in the body can be resolved into two components :

(i) a tangential component $= \alpha r_1$;
(ii) a centripetal component $\omega^2 r_1$ which passes through the axis of rotation O.

As the force on the element acts in the direction of the resultant acceleration, the component forces in the directions of the component accelerations will be :

(i) a tangential force $= m_1 \alpha r_1$ (Force = Mass × Acc.)
(ii) a centripetal force $= m_1 \omega^2 r_1$.

Since the centripetal force component $m_1 \omega^2 r_1$ passes through O it can clearly provide no turning moment about O so that only the tangential force component $m_1 \alpha r_1$ will supply any turning moment about O.

Now the turning moment of mass m_1 about O is given by the product of the force on m_1 and the moment arm length r_1.

Therefore, elemental turning moment $= m_1 \alpha r_1 \times r_1 = \alpha m_1 r_1^2$.

But the body is made up of an infinite number of small particles m_2, m_3, etc., each contributing to the total turning moment on the body.

Then total turning moment or torque $\text{T} = m_1 r_1^2 \alpha + m_2 r_2^2 \alpha + \ldots$

about O.

Because α is constant for all particles we may write

$$\text{T} = \alpha(m_1 r_1^2 + m_2 r_2^2 + \ldots)$$
or $$\text{T} = \alpha \Sigma m r^2,$$

where $\Sigma m r^2$ stands for the summation of all the elemental (mass × radius²) terms.

Mass Moment of Inertia

The value of the term $\Sigma m r^2$, which depends on the radial distribution of all the elemental mass particles about the axis of rotation, is called the

MASS MOMENT OF INERTIA I *

Then about an axis through O the moment of inertia $\Sigma m r^2 = \text{I}_o$.

* 'I' is also used to represent 'second moment of area'. See Chapter VIII.

Hence equation becomes

$$T = I_o\alpha.$$

NOTE.—If torque T is in lb. ft. units, $I_o = \Sigma mr^2$ must be in Slugs ft.² units.

Radius of Gyration

The mass moment of inertia of a body is equal to the sum of all the elemental masses which are distributed throughout the body, multiplied by the square of their own individual radii from the axis of rotation.

Clearly the sum of these distributed masses ($m_1 + m_2 +$ etc.) must equal the total mass of the body M. By imagining this mass M to be concentrated at some particular distance K from the axis of rotation chosen, such that $MK^2 = \Sigma mr^2$, we can express the mass moment of inertia in the more convenient form $I_o = MK_o^2$.

K is called the radius of gyration of the body and care must be taken to specify the axis of rotation with the appropriate subscript.

Centre of Percussion

In many problems we need to establish the line of action of the resultant force F shown in Fig. 98. This resultant force,

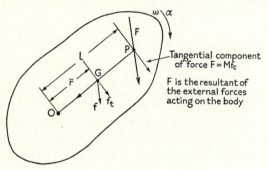

FIG. 98

which is equal to the mass of the body multiplied by the acceleration of the CG, is assumed to pass through some point P on OG produced.

As with the case of a single particle, only the tangential component of F supplies the turning moment about O since the centripetal component passes *through* O.

If f_t is the tangential component of the acceleration of G, the tangential force component through P is $F_t = Mf_t$.

But $f_t = \alpha\bar{r}$, where $\bar{r}$ is the distance of G from the axis of rotation O.

$$\therefore F_t = Mf_t = M\alpha\bar{r}.$$

Now the moment of F_t about O equals the applied torque $T = I_o\alpha$.

$$\therefore F_t \times l = I_o\alpha,$$

where l is the moment arm of F_t about O,

or
$$M\alpha\bar{r} \times l = I_o\alpha.$$

Substituting MK_o^2 for I_o, we get finally

$$M\alpha\bar{r} \times l = MK_o^2\alpha,$$

and
$$l = \frac{K_o^2}{\bar{r}},$$

which gives the distance of the point P from the axis of rotation.

By the parallel axis theorem (Appendix 4) we know that where K_G is the radius of gyration of the body about an axis through its centre of gravity :

Then
$$l = \frac{K_o^2}{\bar{r}} = \frac{K_G^2 + \bar{r}^2}{\bar{r}} = \bar{r} + \frac{K_G^2}{\bar{r}},$$

which indicates that the resultant force F acts—not through G, but beyond the position of G—by an amount equal to $\dfrac{K_G^2}{\bar{r}}$.

As F acts through the point P, it can clearly have no moment *about* P. This point P is then the particular point in the body about which the moment of all the external forces is zero—it is called the Centre of Percussion (cf. Centre of Oscillation in Chapter V, article 34).

Axis of Rotation through G

Most problems at this stage are concerned with rotation about the centre of gravity.

When the axis of rotation O coincides with the centre of

gravity of the body, then $\bar{r}$—the distance between O and G—is zero, or since $K_o{}^2 = K_G{}^2 + \bar{r}^2$, $K_o{}^2$ becomes equal to $K_G{}^2$.

Hence $K_o{}^2$ is replaced by $K_G{}^2$ and I_o becomes I_G in the torque relationship.

Then for rotation about an axis through G we have

$$T = I_G\alpha \ \text{or} \ T = MK_G{}^2\alpha.$$

N.B.—It is much easier to spin a cycle wheel about the spindle axis than about any other axis. This is because the torque required about the axis through the spindle—i.e. through G— is least, because K, the radius of gyration of the wheel, is least about that axis.

Example 35

The total weight of a pair of uniform integral pulleys is 100 lb. and the radius of gyration about the axis of rotation is 13 inches. The effective diameter of each pulley is 12 inches and 40 inches. By means of light strings the large pulley supports a weight of 35 lb. and the small one a weight of 60 lb. If bearing friction is neglected and the weights are released from rest, how far will each move in 3 seconds?

Solution

The direction of motion of the two weights when released from rest can be obtained by taking moments of the weights about the shaft axis. Since the anticlockwise turning effect of the 35-lb. weight acting at a radius of 20 inches is greater than the clockwise-turning effect of the 60-lb. weight acting at a radius of 6 inches, it is evident that the 35-lb. weight moves downwards and the 60-lb. weight upwards.

Let suffixes $(_1)$ and $(_2)$ denote the large and small pulley diameters, respectively.

Then since the angular acceleration α is the same for both pulleys:

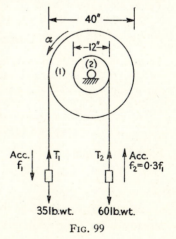

$$\alpha = \frac{f_1}{r_1} = \frac{f_2}{r_2} \ \text{or} \ f_2 = f_1\frac{r_2}{r_1},$$

$$\therefore f_2 = f_1\frac{6}{20} = 0.3f_1.$$

Fig. 99

The torque acting on the pulley system is given by Torque $= I_G \alpha$,

$$\text{i.e. Torque} = \frac{W}{g} k_G^2 \alpha = \frac{100}{32 \cdot 2} \cdot \left(\frac{13}{12}\right)^2 \frac{f_1}{r_1}, \text{ where } r_1 = \frac{20}{12} \text{ ft.}$$

Then   Torque $= 2 \cdot 19 f_1$.

This torque is developed by the difference in the two tensions T_1 and T_2 acting at their respective radii r_1 and r_2 :

$$\text{i.e. } 2 \cdot 19 f_1 = T_1 \times \frac{20}{12} - T_2 \times \frac{6}{12}. \qquad . \qquad . \quad (1)$$

This equation has three unknowns, so two further equations must be obtained for a solution. Those two equations can be deduced from the equations of motion for each weight.

Net force acting on 35-lb. weight $= 35 - T_1$, which acts on the mass $35/g$ slugs to give an acceleration f_1 downwards :

$$\text{i.e. } 35 - T_1 = \frac{35}{32 \cdot 2} f_1. \qquad . \qquad . \qquad . \quad (2)$$

Similarly for the 60-lb. weight for acceleration f_2 upwards :

$$T_2 - 60 = \frac{60}{32 \cdot 2} f_2$$

or substituting for $f_2 = 0 \cdot 3 f_1$

$$T_2 - 60 = \frac{18}{32 \cdot 2} f_1. \qquad . \qquad . \qquad . \quad (3)$$

We thus have three simultaneous equations with the three unknowns T_1, T_2 and f_1 (and hence f_2).

Solving these equations, we get the required accelerations :

$$f_1 = 12 \cdot 5 \text{ ft./sec.}^2 \qquad f_2 = 3 \cdot 75 \text{ ft./sec.}^2$$

To find the distance moved by the weights, we can apply the equation $s = ut + \frac{1}{2} ft.^2$ to each weight with the initial velocity $u = 0$, since weights move from rest.

For 35-lb. weight

$$s = \tfrac{1}{2} 12 \cdot 5 \cdot 3^2 = \textbf{56·25 ft. downwards.}$$

For 60-lb. weight

$$s = \tfrac{1}{2} 3 \cdot 75 \cdot 3^2 = \textbf{16·87 ft. upwards.}$$

28. Energy.—The energy possessed by a body will have already been defined in previous studies as the capacity of the body to *do work* or *overcome resistance* to motion. Two distinct forms of energy concern us in mechanics—namely potential energy and kinetic energy. The potential energy of a body is due to its position, whereas kinetic energy arises solely as a result of the velocity of the body.

The principle of the conservation of energy states that energy can be changed in form but neither created nor destroyed so that in the absence of frictional or other resistances the sum of the potential and kinetic energies remains constant.

Hence any reduction in potential energy is accompanied by a corresponding increase in kinetic energy.

Kinetic Energy of Translation

When a force F acts on a body initially at rest and displaces it in a straight line through a distance S, the work done by F is FS—Fig. 100. The force F will accelerate the body according to the relationship $F = \dfrac{W}{g}f$, where f is the acceleration.

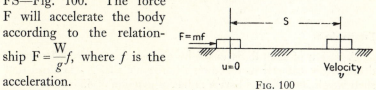

FIG. 100

If the velocity reached in distance S is v then $v^2 = 2fS$ and $S = \dfrac{v^2}{2f}$, since the initial velocity $u = 0$.

Substituting for F and S in the expression Work Done = FS, we get :

$$\text{Work Done} = FS = \frac{W}{g}f \cdot \frac{v^2}{2f} = \frac{Wv^2}{2g}.$$

The term $\dfrac{Wv^2}{2g}$ is the Kinetic Energy possessed by the body weighing W lb., when moving with velocity v ft./sec.

Hence $\qquad$ Linear K.E. $= \frac{1}{2}\dfrac{W}{g}v^2 = \frac{1}{2}Mv^2.$

The units are those of work done, i.e. ft. lb. ; ft. ton, etc.

Kinetic Energy of Rotation

In Fig. 101 the elemental particle of mass m_1 of the body is shown moving with a linear velocity $v = \omega r_1$.

For this velocity the kinetic energy of this elemental mass is given by $\frac{1}{2}$(mass × velocity^2) or

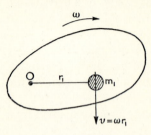

$$\text{K.E. of mass } m_1 = \tfrac{1}{2}m_1(\omega r_1)^2.$$

The total kinetic energy of the body is given by the sum of all such terms, viz. :

$$\text{K.E. of body} = \tfrac{1}{2}m_1\omega^2 r_1{}^2 + \tfrac{1}{2}m_2\omega^2 r_2{}^2 +$$
$$= \tfrac{1}{2}\omega^2\Sigma mr^2.$$

Since ω is the same for all particles.

Fig. 101

But from article 27 $\Sigma mr^2 = \text{I}_o$ the mass moment of inertia about the axis of rotation.

$$\therefore \text{ Kinetic Energy of rotating body} = \tfrac{1}{2}\text{I}_o\omega^2$$

Units. Mass Moment of Inertia in Slugs ft.^{2} ω in rad./sec.

Example 36

A weight of 2500 lb. is raised by means of a rope coiled round a drum 8 feet diameter mounted on a horizontal shaft. The drum and shaft weigh 3000 lb. and have a radius of gyration of 42 inches.

Determine the pull in the rope and the torque required at the shaft to give the weight a constant acceleration of 3 ft./s.^{2} What will be the kinetic energy of the drum after 10 s. ?

(U.L.C.I.)

Solution

The tension in the rope must be sufficient to support the weight and to accelerate it upwards.

$\therefore$ Pull in rope = 2500 + Force to
Acc. weight

$$= 2500 + \frac{2500}{32\cdot2}\,3$$

Pull in rope = 2733 lb.

The torque required at the shaft = Torque to Accelerate drum and shaft + torque due to tension of rope at 4 feet radius.

i.e. Torque = $\text{I}\alpha + 2733 \times 4$

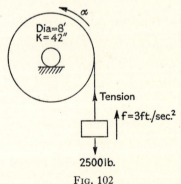

Fig. 102

The mass moment of inertia I

$$= MK^2 = \frac{3000}{32 \cdot 2}\left(\frac{42}{12}\right)^2 = 1140 \text{ Slug ft.}^2$$

Also
$$\alpha = \frac{f}{r} = \tfrac{3}{4} = 0 \cdot 75 \text{ rad./sec.}^2$$

$$\therefore \text{ Torque} = 1140 \times 0 \cdot 75 + 10932$$

Torque = 11788 lb. ft.

The kinetic energy of the drum is given by $\tfrac{1}{2}I\omega^2$.

To determine ω—the angular velocity of the drum after 10 secs.—we have
$$\omega = \alpha t = 0 \cdot 75 \times 10 = 7 \cdot 5 \text{ rad./sec.}$$

Then K.E. of drum after 10 secs.

$$= \tfrac{1}{2}I\omega^2 = \tfrac{1}{2}\, 1140 \times (7 \cdot 5)^2,$$

$$\therefore \text{ K.E.} = 32000 \text{ ft. lb.}$$

Example 37

State the units in which the mass moment of inertia of a flywheel may be measured.

A flywheel weighing 1 ton has a radius of gyration of 2 feet. Calculate (a) the torque necessary to increase the speed of the flywheel from 200 rev./min. to 400 rev./min. in 20 seconds, assuming uniform acceleration, and (b) the average horse power absorbed by the flywheel during this time.   (U.E.I.)

Solution

The mass moment of inertia $I = MK^2$.
For Mass in Slugs Radius of Gyration K in ft.
I is measured in Slug ft.^{2} units.
The equation for the torque required is $T = I\alpha$.

I the mass moment of inertia $= MK^2 = \dfrac{1 \times 2240}{32 \cdot 2} \times (2)^2$.

$$\therefore \; I = 278 \text{ Slug ft.}^2$$

α rad./sec.^{2} is the rate of change of angular velocity,

$$\therefore \; \alpha = \frac{\omega_2 - \omega_1}{t}$$

$$\omega_2 = \frac{2\pi 400}{60} = \frac{40\pi}{3} \text{ rad./sec.} \qquad \omega_1 = \frac{2\pi \cdot 200}{60} = \frac{20\pi}{3} \text{ rad./sec.}$$

$$\omega_2 - \omega_1 = \frac{20\pi}{3} \qquad\qquad \therefore \ \alpha = \frac{20\pi}{3} \Big/ 20 = \pi/3 \text{ rad./sec.}^2$$

Then $T = I\alpha = 278 \times \dfrac{\pi}{3}$ $= \textbf{292 lb. ft.}$

To determine the number of radians moved in 20 seconds we have mean velocity after 20 seconds

$$= \frac{\dfrac{40\pi}{3} + \dfrac{20\pi}{3}}{2}$$

$$= 10\pi \text{ rad./sec.}$$

$\therefore$ Number of radians in 20 seconds = Mean velocity × time
$$= 10\pi \times 20 = 200\pi.$$

Now Work Done by a torque is equal to the torque multiplied by the angle turned through in radians,

$$\therefore \ \text{WD/sec.} = T\theta = 292 \times \frac{200\pi}{20} = 9150 \text{ ft.lb.}$$

$$\text{HP} = \frac{\text{WD/sec.}}{550} = \frac{9150}{550},$$

$$\therefore \ \textbf{HP} = \textbf{16·65.}$$

Alternatively

The change in kinetic energy/sec.

$$= \frac{\frac{1}{2} I(\omega_2{}^2 - \omega_1{}^2)}{20} = 9150 \text{ ft.lb.}$$

Then $\textbf{HP} = \textbf{16·65 as before.}$

K.E. of body having Translation and Rotation

If a body, such as a car wheel, has rotation as well as linear translation, then its total kinetic energy will be the sum of the separate energies for each motion considered separately :

i.e. Total K.E. = K.E. of Translation + K.E. of Rotation.

The proof * of this statement is demonstrated as follows :

Consider the body in Fig. 103 rotating with angular velocity ω about an axis through its CG and let the CG have a linear velocity of v_G. The velocity of any elemental mass m, distance r from the axis of rotation G, is shown as v, and this has been resolved into the two components v_G and ωr. (All particles in a translating body have the same velocity as the CG, so v_G must be the horizontal component of v.)

The component ωr will be at right-angles to the radius r.

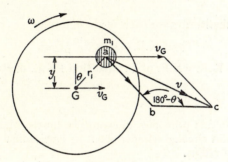

FIG. 103

Now from the geometry of the figure and the triangle abc the velocity v may be expressed in terms of its components by means of the cosine rule.

Hence $\qquad v^2 = v_G^2 + \omega^2 r^2 - 2v_G\omega r \cos (180 - \theta)$.

But $\qquad \cos (180 - \theta) = - \cos \theta,$

$$\therefore \ v^2 = v_G^2 + \omega^2 r^2 + 2v_G\omega r \cos \theta.$$

The kinetic energy of the particle is $\frac{1}{2}mv^2$, and, substituting the value of v^2 given above, we get :

K.E. of particle $= \frac{1}{2}mv_G^2 + \frac{1}{2}m\omega^2 r^2 + \frac{1}{2}m . 2v_G\omega r \cos \theta.$

The total K.E. of the body will be the sum of all the separate energy terms—viz. $\Sigma\frac{1}{2}mv^2$.

Therefore

$$\text{Total K.E.} = \Sigma\frac{1}{2}mv_G^2 + \Sigma\frac{1}{2}m\omega^2 r^2 + \Sigma mv_G\omega . y.$$
Substituting y for $r \cos \theta$.

* Theory may be omitted at this stage.

By removing the values which are constant for all particles this becomes :

$$\text{Total K.E.} = \tfrac{1}{2}v_G\Sigma m + \tfrac{1}{2}\omega^2\Sigma mr^2 + v_G\omega\Sigma my.$$

Now $\Sigma m = \text{M}$ the total mass of the body.

$\Sigma mr^2 = \text{I}_G$ the mass moment of inertia of body about G.

$\Sigma my = 0$ for an axis through G.

This last result follows since Σmy represents the sum of all the first moments of the elements of mass about G, the mass centre.

Hence, substituting for each of these terms,

$$\text{Total K.E.} = \tfrac{1}{2}\text{M}v_G{}^2 + \tfrac{1}{2}\text{I}_G\omega^2.$$

$\therefore$ **Total K.E. = K.E. of Translation + K.E. of Rotation.**

Example 38

Derive from first principles an expression for the Mass Moment of Inertia of a solid circular cylinder about its longitudinal axis in terms of its mass and radius.

A solid cylinder, of weight 5 lb. and radius 2 inches, rolls, without slipping, down a slope inclined at 5° to the horizontal. At a given point on the incline the linear velocity of the cylinder is 2 ft./sec. Determine the distance travelled down the plane when this velocity has increased to 6 ft./sec.

Solution

Consider the solid cylinder of radius R and length L shown in Fig. 104.

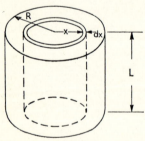

FIG. 104

Let M equal the total mass and m the mass/unit volume.

The volume of any elemental ring of material at radius x of thickness dx is given by $2\pi x \,.\, dx \,.\, \text{L}$.

The mass of the element is $m \,.\, 2\pi x \,.\, dx \,.\, \text{L}$.

Since the mass of this element may be considered concentrated wholly at distance x from the axis we have :

Mass moment of inertia of element about longitudinal axis

$$=(m2\pi x \; dx \; \text{L})x^2.$$

The total mass moment of inertia of the cylinder will be the sum of all such terms,

$$\therefore \text{ Total } I = \int_0^R (m2\pi x \, dx \, L)x^2$$

$$= m2\pi L \int_0^R x^3 dx$$

$$= m2\pi L \cdot \frac{R^4}{4}.$$

Finally   Total $I = mL \cdot \dfrac{\pi R^4}{2}.$

But $\pi R^2 L = $ Volume of cylinder and $m\pi R^2 L = M$ the total mass,

$$\therefore \ I = (m\pi R^2 L)\frac{R^2}{2} = \frac{MR^2}{2}.$$

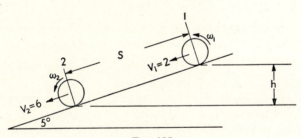

Fig. 105

In travelling a distance S ft. down the incline the cylinder will fall through a vertical height h ft. In doing so it will lose potential energy equal to $Wh = 5h$. Since no slip occurs this energy will be wholly converted into kinetic energy.

Hence the loss of P.E. between (1) and (2) = Gain in the K.E. between (1) and (2).

The gain in K.E. will be partly translational and partly rotational.

Gain in translational K.E. $= \frac{1}{2} M(V_2^2 - V_1^2)$, where V_1 and V_2 represent linear velocities at (1) and (2) respectively,

$$= \frac{1}{2}\frac{5}{32\cdot 2}(6^2 - 2^2)$$

$$= 2\cdot 48 \text{ ft.lb.}$$

Similarly Gain in rotational K.E. $=\frac{1}{2} I_G(\omega_2{}^2 - \omega_1{}^2)$.

$$=\tfrac{1}{2}(\tfrac{1}{2}MR^2)(\omega_2{}^2 - \omega_1{}^2)$$

But $\omega_2 = \dfrac{V_2}{R} = \dfrac{6}{2/12} = 36$ and $\omega_1 = 12$ rad./sec.

$\therefore$ Gain in rotational K.E. $= \tfrac{1}{2} \cdot \tfrac{1}{2} \cdot \dfrac{5}{32\cdot2} \cdot \left(\dfrac{2}{12}\right)^2 (36^2 - 12^2)$

$$= 1\cdot24 \text{ ft.lb.}$$

Then   Loss of P.E. = Gain in Translational K.E.
                                        + Gain in Rotational K.E.

$$5h = 2\cdot48 + 1\cdot24$$

and $h = 0\cdot744$ ft.

but $h = S \sin 5°$.

Distance travelled down incline **S = 8·55 ft.**

29. Linear Impulse and Momentum.—Newton's Second Law, as given in article 20, is really an extension of his original statement that

> the rate of change of *momentum* of a body is proportional to the force producing it and is in the direction of the force.

To deal with the law in this form we must first consider what is meant by the momentum of a body.

The *linear momentum* of a body at any instant is defined as the product of its mass and its velocity at that instant.

Hence a body of mass M moving at a given instant with velocity v will have linear momentum equal to Mv. Linear momentum is a vector quantity with the same sense and direction as the velocity of the body.

We can now express Newton's Law—that force is proportional to the rate of change of linear momentum—in the form :

$$F \propto \frac{d}{dt}(Mv), \qquad \frac{d}{dt} \text{ denoting a rate of change.}$$

Alternatively $$F = K\frac{d}{dt}(Mv),$$

and choosing a consistent set of units, as in article 5, K can be made equal to unity to give :

$$F = \frac{d}{dt}(Mv).$$

For a constant mass M it is evident that only the velocity can change with respect to time so that

$$F = M\frac{dv}{dt}. \qquad \qquad \qquad (1)$$

But $\frac{dv}{dt}$ = the linear acceleration f.

Therefore $\qquad$ F = Mf as previously obtained.

By rearrangement of equation (1) we get : Fdt = Mdv, where dv still represents the change in velocity in a time dt.

If the velocity changes from an initial value u to a final value v in time t, then by integration

$$\int_0^t F dt = M \int_u^v dv.$$

For a constant force F :

$$F \int_0^t dt = M \int_u^v dv,$$

or $\qquad \qquad F t = M(v - u). \qquad \qquad . \qquad (2)$

When F is an impulsive force which is applied for a relatively short interval of time t, the product Ft in this equation is called the *Linear Impulse* of the force F.

M($v - u$) clearly represents the *change* in linear momentum of the mass M in the time t.

Equation (2), therefore, gives the relationship between linear impulse and momentum which may be stated thus :

> the linear impulse of a force which acts on a body during a given interval of time is equal to the change in linear momentum of the body in that time,

i.e. $\qquad$ Impulse = Change in Linear Momentum. $\qquad . \qquad (3)$

Units

The units of momentum and impulse will be the same, as indicated by equation (3).

If the force F is in lb. wt. and the time t is in secs., then impulse and momentum are measured in lb. wt. sec. units.

A.M.—I

Conservation of Linear Momentum

We have shown by equation (3) that impulse equals change in linear momentum. From this relationship it should not be difficult to realise that if there is no impulse there is no change in linear momentum. This fact is expressed by the Principle of the Conservation of Linear Momentum which states that

> if the resultant force (and hence impulse) on a system of bodies is zero in a given direction, the change in linear momentum in that direction is also zero.

Hence, if there is no impulse momentum remains constant.

It is important to realise that the linear momentum only remains unchanged in the directions in which impulse is zero.

Impact

Any collision between bodies which occurs for a short time interval is called an impact.

Consider the impact caused by the collision of the two bodies A and B as depicted in Fig. 106. On impact the linear impulse of A on B is equal and opposite to the linear impulse of B on A. (Newton's Third Law.)

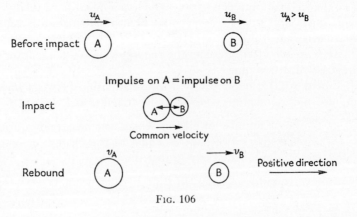

Fig. 106

It follows, then, that the change in momentum of the two bodies is also equal and opposite, and therefore, the momentum in the direction of the impulse before impact will equal the momentum in that direction after impact.

If u and v represent velocities before and after impact respectively, we have, by the principle of conservation of momentum,

momentum before impact = momentum after impact

$$M_A u_A + M_B u_B = M_A v_A + M_B v_B. \qquad . \qquad (4)$$

When using this equation due regard must be made to direction and since the velocities of both bodies after impact may be unknown they should both be assumed positive. Velocities which are then evaluated as negative will indicate that those bodies move in the direction opposite to that selected as positive.

During the short period of impact the bodies move together with a common velocity and then rebound with the velocities v_A and v_B.

It is found by experiment that the ratio of the relative velocity of the two bodies *after* impact to their relative velocity *before* impact is a constant quantity which depends on the elastic properties of the two bodies.

This ratio is called the coefficient of restitution e.

$$e = -\left(\frac{\text{Relative velocity after impact}}{\text{Relative velocity before impact}}\right) = -\left(\frac{v_A - v_B}{u_A - u_B}\right). \qquad (5)$$

The negative sign is introduced into this equation because the relative velocities before and after impact are always of opposite sense. The value of 'e' is then a positive quantity.

Equation (5), together with that obtained by applying the conservation of momentum principle (equation 4), will be sufficient to determine the unknown velocities after impact.

For perfectly elastic bodies e is unity, whereas for completely inelastic bodies $e = $ zero. For this latter case the bodies would not rebound but would move together after impact.

Loss of Kinetic Energy due to Impact

Except for the ideal case of perfectly elastic impact, $e = 1$, there will always be some loss of energy due to impact. For this reason the kinetic energy after impact will always be LESS than the kinetic energy of the bodies before impact.

For the case when $e = 0$—i.e. for inelastic impact—the loss of kinetic energy will be a maximum.

Example 39

A block weighing 20 lb. is moving along a smooth horizontal plane with a velocity of 50 ft./sec. to the left when it is struck centrally by a bullet weighing 1 oz. which passes right through it. The velocity of the bullet changes from 2400 ft./sec. to the right before impact to 400 ft./sec. to the right after impact.

Determine : (a) the velocity of the block just after impact ; (b) the linear impulse on the block due to impact.

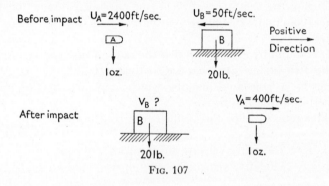

Fig. 107

(a) For a smooth horizontal plane friction forces are negligible. Then Momentum before impact = Momentum after impact

$$M_A U_A + M_B U_B = M_A V_A + M_B V_B.$$

$$\frac{1}{16 \cdot 32\cdot2} \cdot 2400 + \frac{20}{32\cdot2}(-50) = \frac{1 \cdot 400}{16 \cdot 32\cdot2} + \frac{20}{32\cdot2} V_B.$$

Multiplying throughout by 32·2 :

$$150 - 1000 = 25 + 20 V_B.$$

$$\therefore V_B = -43\cdot75 \text{ ft./sec.}$$

The negative sign indicates that the block continues to move to the left.

(b) The linear impulse ON the block is due to the change in momentum of the bullet. The velocity of the bullet changes from 2400 ft./sec. to 400 ft./sec. in the positive direction.

Then   Impulse = Change in Momentum of bullet

$$Ft = M_A(U_A - V_A),$$

$$\text{Impulse} = \frac{1}{16 \cdot 32 \cdot 2}(2400 - 400) = +3 \cdot 88 \text{ lb. sec.}$$

NB.—Impulse ON the block is in the positive direction.

Oblique Impact

It was stated in the preceding theory that for equal and opposite impulses the momentum of two bodies remains unchanged *in the direction of the impulse*.

Hence, for an oblique collision as illustrated in Fig. 108, the linear momentum will remain unchanged along the line of centres at impact—i.e. in the *x* direction—and therefore, only component velocities in this direction must be employed in the momentum equation. This also applies to the equation giving *e* in terms of the relative velocities.

For the impact shown in Fig. 108 it should be observed that there is no force acting on either of the

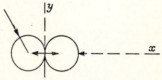

Momentum equation applied in direction of impulse using component velocities in this direction

FIG. 108

bodies at right-angles to the direction of the impulse so that velocity components in the *y* direction will remain unchanged by the impact.

The following example illustrates the method of solution for oblique impact.

Example 40

A smooth ball A, weighing 1 lb., collides with a similar ball B, weighing 2 lb., as shown in the diagram. The velocity of A before impact is inclined at 60° to the line joining the centres of the two balls at the moment of impact, whilst the velocity of B before impact is along this line. If the coefficient of restitution for the bodies is 0·8 determine :

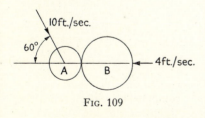

FIG. 109

(*a*) the velocity of the two bodies after impact ;
(*b*) the loss of kinetic energy due to impact.

Solution

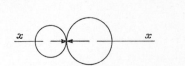

| Impact forces equal and opposite along *xx*. ∴ Momentum unchanged in this direction. | For body A *y* component $5\sqrt{3}$ remains unchanged by impact. | For body B velocity in *y* direction before impact equals velocity in *y* direction after impact equals zero. |

Fig. 110

Neglect any rotational effects.

Let V_A and V_B represent component velocities in the *x* direction after impact.

Then applying principle of conservation of momentum in *x* direction :

$$\frac{1}{32 \cdot 2} \times (10 \cos 60°) + \frac{2}{32 \cdot 2} \times (-4) = \frac{1}{32 \cdot 2}V_A + \frac{2}{32 \cdot 2}V_B.$$

V_A and V_B initially considered positive.

This becomes $5 - 8 = V_A + 2V_B,$

or $V_A + 2V_B = -3.$.   .   . (1)

Also $e = -\dfrac{\text{(Relative velocity after impact)}}{\text{Relative velocity before impact}},$

In this equation we must use the velocity components along the line of impact *xx*.

Thus $0 \cdot 8 = -\dfrac{(V_A - V_B)}{(10 \cos 60° - (-4))} = \dfrac{-(V_A - V_B)}{9}.$

$$\therefore V_A - V_B = -7 \cdot 2. \qquad . \qquad . \qquad (2)$$

Solving the simultaneous equations (1) and (2) we get :

$$V_B = +1 \cdot 4 \text{ ft./sec.} \qquad V_A = -5 \cdot 8 \text{ ft./sec.}$$

These values give the velocity components in the *x* direction.

Since the velocity of B has no component in the *y* direction, $V_B = 1 \cdot 4$ ft./sec. represents the final velocity of B after impact, i.e. B's velocity is REVERSED and it moves to the right with a velocity of $1 \cdot 4$ ft./sec.

The final velocity of A is the resultant of the component—5·8 ft./sec.—in the x direction and the *unchanged* component—$5\sqrt{3}$ ft./sec.—in the y direction.

Final velocity of A $= \sqrt{(5\cdot8)^2 + (5\sqrt{3})^2}$

$= \mathbf{10\cdot4}$ **ft./sec.**

The direction of this velocity with the line of impact is given by :

$$\tan\theta = \frac{5\sqrt{3}}{5\cdot8} \text{ or } \theta = \tan^{-1} 1\cdot493 = 56\cdot2°.$$

K.E. of bodies before impact $= \frac{1}{2} \cdot \frac{1}{32\cdot2}(10)^2 + \frac{1}{2} \cdot \frac{2}{32\cdot2}(4)^2 = 2\cdot046$ ft.lb.

K.E. of bodies after impact $= \frac{1}{2} \cdot \frac{1}{32\cdot2}(10\cdot4)^2 + \frac{1}{2} \cdot \frac{2}{32\cdot2} \cdot (1\cdot4)^2$

$= 1\cdot741$ ft.lb.

Hence Loss of K.E. due to impact $= 2\cdot046 - 1\cdot741 \backsimeq \mathbf{0\cdot30}$ **ft.lb.**

Impact of a Fluid Jet

The determination of the forces developed when a jet of fluid is deflected by a fixed blade provides a further application of the principles of linear impulse and momentum.

When dealing with this type of problem we are usually concerned with the rate of flow or discharge per unit time, and in such cases the impulse-momentum equation :

$Ft = M(u - v)$ is often expressed in the form

$$F = \frac{M}{t}(u - v).$$

In this rearrangement $\frac{M}{t}$ represents the mass flowing/unit time and the equation can now be stated as,

Force due to impact = Mass flowing/sec. × Change of velocity
(in direction of force).

Example 41

A jet of water 1 inch diameter is discharged perpendicularly to a fixed plate with a velocity of 100 ft./sec. What is the force on the plate ?

Solution

When the jet strikes the plate it spreads out radially and, therefore, the velocity of the jet *perpendicular* to the plate is completely
destroyed.

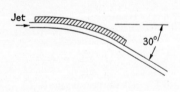

The change in velocity of the jet
$(u - v) = 100 - 0 = 100$ ft./sec.

In one second the volume of water
displaced = Area of jet × velocity of jet,

$$= \frac{\pi}{4} \cdot \left(\frac{1}{12}\right)^2 \times 100 \text{ ft.}^3$$

$$= 0{\cdot}545 \text{ ft.}^3$$

Momentum perpendicular
to the plate is destroyed

Fig. 111

Then mass flowing/second

$$= 0{\cdot}545 \times \frac{62{\cdot}4}{32{\cdot}2}$$

$$= \textbf{1}{\cdot}\textbf{05 Slugs.}$$

Force on plate

= Mass flowing/sec.
                × Change of velocity

$$= 1{\cdot}05 \times 100,$$

∴ **Force on Plate** = **105 lb.**

Example 42

A jet of water moving at 60 ft./sec. is deflected through an angle of
30° by a fixed curved blade. The jet discharges 10 lb./sec., which
enters the blade tangentially. If friction between the water and the
blade is negligible, find the force of the water on the blade. What
would be the value of this force if the jet was deflected through an
angle of 150° ?

Solution

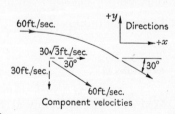

Fig. 112

In the absence of frictional resistance the velocity of the water relative to the blade will remain unchanged. Therefore, velocity at exit = 60 ft./sec. The actual force of the water on the blade can be determined from the component forces acting in the x and y directions.

Component force on blade in x direction $F_x = \dfrac{M}{t}(U_x - V_x)$.

Component force on blade in y direction $F_y = \dfrac{M}{t}(U_y - V_y)$.

To determine F_x

The initial velocity at entry in the positive x direction $U_x = 60$ ft./sec., and the final velocity at exit in same direction $V_x = 30\sqrt{3}$ ft./sec.

$$\therefore \text{ change in velocity} = (60 - 30\sqrt{3}) = 8 \text{ ft./sec.}$$

Then $$F_x = \frac{10}{32 \cdot 2} \times 8 = +2 \cdot 49 \text{ lb.}$$

To determine F_y

Initial velocity at entry in *positive* direction $U_y = 0$. Final velocity at exit in same direction $V_y = -30$ ft./sec.

$$\therefore \text{ change in velocity} = (0 - (-30)) = +30 \text{ ft./sec.}$$

Then $$F_y = \frac{10}{32 \cdot 2} \times 30 = +9 \cdot 32 \text{ lb.}$$

The resultant force on the blade $F = \sqrt{F_x{}^2 + F_y{}^2} = \sqrt{(2 \cdot 49)^2 + (9 \cdot 32)^2}$

$$= 9 \cdot 65 \text{ lb.}$$

Jet deflected through 150°

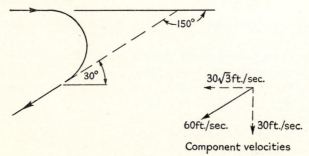

Component velocities

FIG. 113

A comparison of the component velocity diagrams in Figs. 112 and 113 should indicate that whilst the x direction component is completely reversed the y component remains—30 ft./sec.

Then for F_x Change in velocity $= (60 - (-30\ 3\sqrt{3})) = 112$ ft./sec.

$$\therefore\ F_x = \frac{10}{32\cdot2} \times 112 = +34\cdot8\ \text{lb.}$$

$$F_y = +9\cdot32\ \text{lb. (as before)}$$

Resultant force on blade $F = \sqrt{F_x{}^2 + F_y{}^2}$

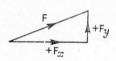

$$= \sqrt{(34\cdot8)^2 + (9\cdot32)^2}$$

$$= \textbf{36 lb.}$$

Fig. 114

It should be observed that with the notation employed in this example the component forces *on* the blade act in the same direction as the component changes in velocity. The force on the jet is equal to F in the opposite direction.

30. Angular Impulse and Momentum.—*Angular impulse* is defined as the *moment* of linear impulse. The moment of the linear impulse of a force on any body about any given axis is called the angular impulse of the force about that axis.

If r is the distance from a given axis at which a constant force F acts for a time interval t, then since Ft is the linear impulse of the force the product $Ft \times r$ equals the angular impulse or moment of impulse :

i.e. Angular impulse $= Ft \times r$.

But the product Fr = Torque or Turning Moment T,

$\therefore$ Angular Impulse $= Tt$.

Angular Momentum is defined as the moment of linear momentum.

The linear momentum of any particle of mass m in a body, moving at a given instant with velocity v, is mv. The moment of the linear momentum with respect to an axis through O (Fig. 115) is linear momentum multiplied by radius.

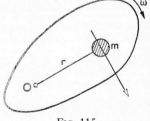

Fig. 115

Then moment of linear momentum = $mv \,.\, \times r$.

Now any body rotating with angular velocity ω will be made up of an infinite number of small masses such as m,

$$\therefore \text{ Moment of momentum for whole body} = \int mvr.$$

But $v = \omega r$ $\qquad \therefore$ Moment of momentum $= \int m\omega r^2,$

$$\text{which for constant } \omega = \omega \int mr^2.$$

Since $\int mr^2$ is the Mass moment of inertia I, we have Moment of Momentum = $I\omega$,

or $\qquad\qquad$ **Angular Momentum** $= I\omega$

For a *change* in angular velocity $(\omega_2 - \omega_1)$ the *Change* in Angular Momentum $= I(\omega_2 - \omega_1)$.

The relationship between angular momentum and angular impulse is similar to that developed for the linear case :

$$\therefore \text{ Angular Impulse} = \text{Change in Angular Momentum}$$

or $\qquad\qquad$ $Tt = I\ (\omega_2 - \omega_1).$

The conservation of momentum can then be stated for the angular case as follows :

> If the resultant externally applied moment or torque on any system about a fixed axis is zero the angular momentum remains constant.

As in the linear case, the momentum of each part of a system may change but the total angular momentum remains constant as illustrated in the following example.

Example 43

A and B are two separate clutch plates. A weighs 96 lb., has radius of gyration 6 inches and rotates at 600 rev./min. B weighs 72 lb., has radius of gyration 4 inches and is stationary. When the clutch plates are engaged slipping ceases to occur after 0·20 sec.
Determine :

(a) the common angular velocity of the plates ;
(b) the loss of kinetic energy ;
(c) the angular acceleration of plate B ;

(d) the torque transmitted by the clutch assuming that this is constant during the 0·20 sec. period;

(e) the angle in radians through which the one face of the clutch slips, relatively to the other face.

Solution

(a) There is no externally applied torque on the system and, therefore, the angular momentum remains constant.

Hence Angular momentum before impact = Angular momentum after impact.

Momentum before impact is that due to plate A only as B is stationary.

After impact both plates move with common velocity ω_C.

Then we get $I_A\omega_A + O = (I_A + I_B)\omega_C$,

$$\left[\frac{96}{32\cdot2} \times \left(\frac{6}{12}\right)^2\right] \cdot \frac{2\pi600}{60} = \left[\frac{96}{32\cdot2} \cdot \left(\frac{6}{12}\right)^2 + \frac{72}{32\cdot2} \cdot \left(\frac{4}{12}\right)^2\right] \cdot \frac{2\pi N_C}{60}.$$

Cancelling common terms:

$$(96 \times \tfrac{1}{4})600 = [96 \times \tfrac{1}{4} + 72 \times \tfrac{1}{9}]N_C,$$

$$\therefore N_C = \textbf{450 rev./min.}$$

(b) Loss of Kinetic Energy = K.E. before impact − K.E. after impact.

K.E. before impact $= \tfrac{1}{2}I_A\omega_A^2 + O$ (Since B is at rest.)

$$= \tfrac{1}{2} \cdot \frac{96}{32\cdot2} \cdot \left(\frac{6}{12}\right)^2 \cdot \left(\frac{2\pi600}{60}\right)^2,$$

$\therefore$ K.E. before impact = 1481 ft.lb.

K.E. after impact $= (\tfrac{1}{2}I_A + \tfrac{1}{2}I_B)\omega_C^2$

$$= \left(\tfrac{1}{2} \cdot \frac{96}{32\cdot2} \cdot \left(\frac{6}{12}\right)^2 + \tfrac{1}{2} \cdot \frac{72}{32\cdot2} \cdot \left(\frac{4}{12}\right)^2\right)\left(\frac{2\pi450}{60}\right)^2$$

$$= 1110 \text{ ft.lb.}$$

$\therefore$ **Loss of K.E.** = 1481 − 1110 = **371 ft.lb.**

(c) Plate B is accelerated from rest to the common velocity 450 rev./min. in 0·2 sec.

Using $\omega_C = \omega_B + \alpha_B t$, where $\omega_B = O$, we get:

$$\frac{2\pi450}{60} = O + \alpha_B \times 0\cdot2,$$

$$\therefore \alpha_B = \textbf{236 rad./sec.}^2$$

(d) The torque transmitted by the clutch is given by $T = I_B \alpha_B$.

$$\therefore T = \frac{72}{32 \cdot 2} \cdot \left(\frac{4}{12}\right)^2 . 236$$

$$T = 58 \cdot 7 \text{ lb.ft.}$$

Since the impulse on both plates is equal and opposite this torque could equally be obtained from $T = I_A \alpha_A$, where α_A is the *deceleration* of plate A.

Plate A is decelerated from 600 rev./min. to 450 rev./min. in 0·2 sec.

$$\text{i.e. } \omega_C = \omega_A - \alpha_A t$$

or

$$\frac{450 \cdot 2\pi}{60} = \frac{600 \cdot 2\pi}{60} - \alpha_A 0 \cdot 2$$

$$\alpha_A = 78 \cdot 5 \text{ rad./sec.}^2$$

Then

$$T = I_A \alpha_A = \frac{96}{32 \cdot 2} \cdot \left(\frac{6}{12}\right)^2 \times 78 \cdot 5 = 58 \cdot 7 \text{ lb.ft. as before.}$$

(e) The energy is lost during the period of slip and the work done by the torque during this period $= T\theta$ where θ in radians is the *relative* angle of slip.

Then

$$T\theta = \text{Loss of K.E.}$$
$$58 \cdot 7\theta = 371$$

$$\therefore \theta = 6 \cdot 33 \text{ radians.}$$

The following alternative method could be used to give the same result for θ, although it is much longer.

Plate B.—This is accelerated from rest to 450 rev./min, with an angular acceleration of 236 rad./sec.[2]

Using $\omega_C^2 = \omega_B^2 + 2\alpha\theta_B$, where θ_B is the angle turned through by B and $\omega_B = 0$.

Then

$$\left(2\pi \frac{450}{60}\right)^2 = 2 \times 236 \times \theta_B.$$

$$\theta_B = 4 \cdot 7 \text{ radians.}$$

Similarly for Plate A, which is decelerated from 600 rev./min. to 450 rev./min., with a deceleration of 78·5 rad./sec.[2]

$$\omega_C^2 = \omega_A^2 - 2a\theta_A.$$
$$\theta_A = 11 \cdot 03 \text{ radians.}$$

Now during slipping both plates are moving in the *same* direction so that the angle through which one face of the clutch *slips relative* to the other is given by $\theta_A - \theta_B = 11 \cdot 03 - 4 \cdot 70 = 6 \cdot 33$ radians as before.

EXERCISES IV

1. Define the term 'acceleration'.

If a body is moving in a straight line with uniform acceleration f, obtain an expression for the distance travelled S whilst the velocity changes from V_1 to V_2.

After the first operation in a power press each part weighs 0·1 lb., and for the next operation the parts fall singly and vertically from rest onto a platform through a distance of 3 inches. If the vertical motion is arrested in 0·055 second, calculate the average force each part exerts on the platform. I.Prod.E.

2. Newton's Second Law of Motion states: 'Force is proportional to rate of change of momentum'. Derive the expression connecting force, mass and acceleration, stating the units used for each quantity.

The draw-bar pull of a locomotive is 15 tons. Assuming a rolling resistance of 12 lb./ton determine the slope of the steepest gradient that could be climbed with an acceleration of 0·5 ft./sec.², if the weight of the locomotive and train is 500 tons. What power is the engine developing when a speed of 9 m.p.h. is reached? N.C.T.E.C.

3. A load of 200 lb. is drawn up an incline at 12° to the horizontal for a distance of 100 feet, the coefficient of friction being 0·3. If the force pulling parallel to the slope is 112 lb. determine:

(a) the acceleration of the load;
(b) the amount of work done against friction.

4. A lorry having a total weight of 4 tons is hauled up an incline of 1 in 4 (measured as a sine) by means of a long rope which passes round a drum at the top of the incline. The drum is 3·5 feet diameter; radius of gyration 1·5 feet and weight 2 tons. Each of the four lorry wheels weighs 240 lb., is 30 inches diameter and has radius of gyration of 12 inches.

The coefficient of rolling friction between the tyres and the incline is 0·2; also there is an opposing wind resistance of 20 lb. per ton. If a constant torque of 7350 lb. ft. is applied to the drum shaft, calculate the distance the lorry travels up the incline whilst its speed increases from 5 to 10 ft./sec.

5. A locomotive weighing 70 tons is attached to a train of weight 250 tons, and is drawing it up an incline of 1 in 160 against a frictional resistance of 15 lb. per ton for both engine and train. The speed is 25 m./hour at a particular instant but is increasing at the rate of 1 m./hour in 11 seconds. Find the pull in tons weight in the coupling between the engine and the train and the H.P. at which the engine is working.

6. The cage of a lift weighs 800 lb., and is drawn upward by a wire rope. The cage starts from rest and initially the pull on the rope is 1400 lb. The pull decreases uniformly with distance ascended until at a height of 80 feet the pull is 600 lb. Calculate the velocity of the lift at this height. Calculate also the greatest velocity of the lift and the height at which it occurs.

7. The rotor of a steam turbine weighs 40 lb., and has a radius of gyration of $1\frac{1}{2}$ inches. The dangerous speed range for this rotor is from 20,000 rev./min., to 18,000 rev./min. Calculate the smallest braking torque necessary to ensure that the turbine does not take longer than 2 seconds to pass through this speed range when the steam is shut off and the average horse power absorbed during this period. If the blades weigh $\frac{3}{4}$ ounce each and their centres of gravity are on a circle of 6-inch diameter calculate the centripetal force on each blade at 30,000 rev./min.

8. A small body rests on a 14-inch diameter rotating table at a distance of 3 inches from the axis of rotation. If the coefficient of friction between the body and the table is 0·25, at what speed will the body begin to slip ? If slipping can only take place in a radial direction, at what angle will the body leave the table ?

9. Derive an expression for the 'Centripetal Acceleration' of a particle rotating in a circular path with uniform angular velocity.

A shaft, 30 inches long, is simply supported at each end and has a flywheel weighing 96 lb. fixed to it at a distance of 10 inches from the left-hand support. The centre of gravity of the flywheel is 0·25 inch from the centre of the shaft.

(*a*) If the shaft is rotated at 20 rev./min., what are the maximum and minimum reactions at each support ? (*b*) At what speed must the shaft be rotated so that the reactions at the supports could be zero at some stage of revolution of the shaft ?

10. A light spring is threaded on a rod and secured to one end of it. The other end of the spring is attached to a 10-lb. annular mass capable of sliding without friction along the rod. When stationary the centre of gravity of the weight is 1 foot from the end of the rod to which the spring is attached. The rod is then rotated about this end in a horizontal plane at 40 r.p.m.

If the stiffness of the spring is 10 lb./ft., find the radius of rotation of the centre of gravity of the 10-lb. weight. In this position the weight is prevented from further outward movement by a stop whilst the speed is increasing to 80 rev./min. The stop is then released. Find the initial acceleration of the weight.

11. Define the term 'centripetal acceleration'.

The road surface of a hump-backed bridge has a radius of curvature of 27 feet. Calculate the greatest speed at which a car can cross the bridge without the wheels leaving the surface, if the centre of gravity of the car is 3 feet above the ground. U.E.I.

12. A four-wheeled, two-axled vehicle travels at a uniform speed of 45 miles/h. round a curve of 800 feet radius. The distance between the wheel tracks is 5·5 feet, the centre of gravity of the vehicle is on the vertical through the centre of the wheel base and 5 feet above the ground. The total weight of the vehicle is 20 tons.

Determine (a) the vertical pressure on each of the inner and outer wheels, (b) the maximum speed at which the vehicle can negotiate the curve without overturning. U.L.C.I.

13. Deduce an expression for the centripetal force required to induce a body of weight W moving with velocity v to move in a circular path of radius r.

A motor cyclist rounds a curve of 500 feet radius at a speed of 30 mile/h. : (a) at what angle must he lean ? ; (b) what is the least coefficient of friction between the wheels and ground required to prevent side slip ? U.E.I.

14. Explain with the aid of a diagram the relationship between centripetal and centrifugal force.

Calculate the greatest speed at which a railway truck can move round a curve of 500 feet mean radius without overturning, if the centre of gravity of the truck is 6 feet above the rails and the gauge of the track is $4\frac{1}{2}$ feet. Assume the track to be horizontal. What fraction of the weight is the horizontal force on the wheel flanges at this speed ? U.E.I.

15. A bend is in the form of an arc of a circle of radius 250 feet, and is banked at an angle $\tan^{-1} 1/5$. Find the speed at which a vehicle may travel round this bend without side thrust on the tyres, and find the speed of impending skidding if the coefficient of friction between tyres and road is $3/5$.

16. A train weighing 200 tons travels round a curved track of 1400 feet radius. The centre of gravity of the train is 6 feet above the rails, which are 4 feet 6 inches apart. Working from first principles, determine the super-elevation of the outer rail if side thrust is to be eliminated at the maximum allowable speed of 30 m.p.h. You may assume the expression for centripetal acceleration.

Explain with the aid of a diagram the forces which are acting on the engine whilst travelling round the bend, giving their values. N.C.T.E.C.

17. Deduce from first principles the formula giving the relationship between the torque and angular acceleration for a rotating drum. A cage weighing 2 tons is raised by means of a vertical rope which passes round a horizontal drum of effective diameter, 4·5 feet ; radius of gyration, 2 feet ; weight of drum, 4 tons. If the cage is to be raised with a uniform acceleration of 1·2 ft./sec.², determine the torque required at the drum shaft.

18. Explain what is meant by 'moment of inertia' and show that the kinetic energy of a body rotating about a fixed axis is given by $\frac{1}{2}I\omega^2$ where I is the moment of inertia and ω the angular velocity.

A rope hanging over a pulley of radius 2 feet has attached to it at one end a weight of 24 lb. and at the other end a weight of 20 lb. Initially, motion is prevented by holding one of the weights. If the moment of inertia of the pulley about its axis of rotation is 336 lb. ft.², find the velocity of the weights after they have moved a distance of 4 feet from rest. Assume no slip between rope and pulley and that the pulley runs on frictionless bearings.

19. A weight of 10 lb. hangs vertically and is attached to a cord wound round the horizontal axle of a flywheel. The flywheel weighs 50 lb. and has a radius of gyration about its axis of 6 inches, and the axle is 2 inches diameter. If the weight is released, and in falling commences to turn the flywheel, calculate (*a*) the acceleration of the weight, (*b*) the time taken for the weight to fall vertically from rest through a distance of 6 feet, (*c*) the tension in the cord. Neglect friction at the axle bearings. U.L.C.I.

20. What is the function of a flywheel ?

A weight of 40 lb. attached to a cord which is wrapped round the 2 inches diameter spindle of a flywheel descends and thereby causes the wheel to rotate. If the weight descends 6 feet in 10 seconds and the friction of the bearing is equivalent to a force of 3 lb. at the circumference of the spindle, find the moment of inertia of the flywheel. If it weighs 212 lb., what is its radius of gyration ? N.C.T.E.C.

21. In a flywheel experiment a weight of 8 lb. is fastened to a cord which is attached to a peg on the axle. The axle is 1·875 inches diameter and the cord is 0·125 inch diameter. The 8-lb. weight descends 5 feet in 5 seconds—the 8 lb. weight drops off and the flywheel does a further 100 revolutions before coming to rest. If the flywheel weighs 64 lb., calculate (*a*) the radius of gyration of the flywheel, (*b*) the total friction torque at the bearings.

22. A flywheel weighs 10 tons and has a radius of gyration of 3 feet. Find the time taken for the flywheel to reach a speed of 300 r.p.m. from rest, if friction is neglected and the wheel is subjected to a constant torque of 500 lb. ft. Determine the number of revolutions the wheel would make in coming to rest from 300 r.p.m. under the action of a friction torque of 120 lb. ft. N.C.T.E.C.

A.M.—K

23. Explain the functions of a flywheel.

(*a*) A flywheel weighing 125 lb. and having a radius of gyration of 4·25 inches runs at a uniform speed of 500 rev./min. Calculate its store of kinetic energy.

(*b*) If a machine absorbs 200 ft.lb. of this energy, calculate the new flywheel speed.

(*c*) Find the mean retardation of the wheel and the number of revolutions it makes if the speed change in (*b*) takes place in 5 seconds.

I.Prod.E.

24. The rim of a flywheel is 48 inches external diameter, 40 inches internal diameter and 6 inches wide. It is connected to the hub by a solid web.

(*a*) If the specific weight of the material is 0·25 lb./in³, determine the moment of inertia of the flywheel, stating clearly the units employed. Work from first principles or prove any formula used. (The effect of the web and hub is to be disregarded.)

(*b*) Calculate the energy in ft.lb. stored in the flywheel when rotating at 600 rev./min.

U.E.I.

25. A single-action blanking press has a flywheel 3 feet diameter weighing 1800 lb. and running at 300 rev./min. During a certain blanking operation through $\frac{1}{4}$-inch steel plate a force of 100 tons is required on the punch.

(*a*) Calculate the new speed of the flywheel just after the operation is complete assuming that no energy is supplied from the motor during the operation.

(*b*) If the flywheel must now be run up to its original speed in a time of 5 seconds, what torque must be supplied to it from the motor ?

I.Prod.E.

26. A flywheel of weight 50 lb. is mounted on a horizontal axle 2 inches diameter. A cord is wound round the axle and a weight of 3 lb. is attached to the free end. The flywheel is held at rest and then released and the weight takes 15 seconds to fall through a height of 5 ft. Assuming that the frictional torque on the axle remains constant and is equal to $\frac{1}{2}$ lb. in. determine :

(*a*) the angular acceleration of the flywheel ;
(*b*) the radius of gyration of the flywheel.

27. The flywheel of a power press weighs 500 lb. and has a radius of gyration of 8·5 inches. The flywheel is rotating at 500 rev./min. when the ram is operated. The press is blanking material 0·5 inch thick, the average force on the ram being 25 tons. Calculate the reduction in speed of the flywheel during the blanking process.

28. A lift cage, with contents, weighs 3200 lb. and is raised by means of a wire rope passing round a drum of 3 feet diameter. The drum, which weighs 480 lb. and has a radius of gyration of 1 foot, is driven by an electric motor which gives a maximum torque of 6000 lb. ft. at the drum. If there is a resisting torque at the drum amounting to 5 per cent of the applied torque, calculate the maximum acceleration of the cage, and the tension in the wire rope.

29. A machine punching 1·5-inch diameter holes in 0·5-inch thick plate does 16 in. ton of work per square inch of sheared area. The flywheel of this machine weighs 1·6 tons and has a radius of gyration of 2 feet.

(a) If the flywheel normally runs at 90 rev./min, at no load, determine the fall in the speed of the flywheel due to punching one hole, assuming that the energy required for punching the hole is supplied entirely by the flywheel.

(b) If this flywheel is rotating in two bearings, each 6 inches diameter, and the coefficient of friction is 0·012, calculate the number of revolutions performed by the flywheel in coming to rest (due to bearing friction only) from a speed of 90 rev./min., assuming that the power supply has been removed.

30. Define Potential and Kinetic Energy.

A pile-driver hammer weighing 2000 lb. falls a distance of 2 feet onto an inelastic pile weighing 1000 lb. and drives it a distance of 2 inches into the ground against uniform resistance. Calculate :

    (a) the velocity with which the pile begins to move ;
    (b) the time during which it is moving ;
    (c) the value of the ground resistance.

31. State the Principle of the Conservation of Momentum.

A pile driver of mass 5 cwt. falls through a vertical height of 26 feet and strikes a pile of mass 15 cwt. which is driven into the ground a distance of 3 inches. Assuming that after the blow the pile and driver move together and that the ground resistance is uniform, determine (a) the average ground resistance, and (b) the loss of kinetic energy due to impact.

32. A and B are two waggons in the same straight line. A weighs 4 tons and moves at 16 ft./sec. B weighs 8 tons and moves at 2 ft./sec. in the opposite direction to A. If the two waggons collide head-on and then begin to move with a common velocity, determine (a) the common velocity, (b) the loss of kinetic energy due to impact, (c) the distance the two waggons move together immediately after impact if there is an opposing rail resistance of 14 lb./ton.

33. A jet of water 2 inches diameter having a velocity of 60 ft./sec. impinges tangentially on a curved vane which deflects the jet through an angle of 120°. Calculate the magnitude of the resultant force on the vane.

34. A circular jet of water delivers 2 cubic feet per second with a velocity of 80 ft./sec. and impinges tangentially on a single vane moving in the direction of the jet with a velocity of 40 ft./sec. The vane is so shaped that if stationary it would deflect the jet through an angle of 45°. Determine :

(i) the angle through which the moving vane deflects the jet ;

(ii) the force on the vane (*a*) in the direction of motion ; (*b*) at right angles to the direction of motion.

35. The two halves of a discharged friction clutch are mounted on their independent co-axial shafts (neither being driven from an external source), one element A and its shaft running freely at 420 rev./min. and the other element B meanwhile rotating freely at 300 rev./min. in the opposite direction. The clutch is now engaged. Find the common speed of revolution to which the clutch settles down. A weighs 30 lb. and has a radius of gyration of 6 inches, B weighs 42 lb. and has a radius of gyration of $4\frac{1}{2}$ inches. How much kinetic energy is lost in the process ?

Chapter V

SIMPLE HARMONIC MOTION

MOTION which repeats itself after a time interval is said to be periodic.

A simple example of this kind of motion is obtained by holding a scale rule over the edge of a table and tapping its overhanging end. If the disturbing force is not re-applied, the end of the rule is said to vibrate or oscillate 'freely' and it is the type of repeated to and fro motion of the end of the rule which we now have to analyse.

With such a periodic motion the vibrations do not continue indefinitely, but are eventually 'damped' out by the internal resistance of the body and the external resistance of the air. However, as these damping effects are often very small, we shall at this stage neglect them and confine the analysis to those common types of free vibration which approximate to simple harmonic motion.

31. Simple Harmonic Motion.—Simple Harmonic Motion is defined as the motion of a body whose acceleration is always directed *towards* a fixed point in its path and is proportional to its *distance* from that point.

Consider an imaginary point P^1 which moves around a circle of radius r with constant angular velocity ω rad./sec.—Fig. 116. It will be shown that the oscillation of the point P—the projection of P^1 on the diameter AB—takes place with Simple Harmonic Motion *across the diameter AB*.

The Auxiliary Circle

The circle of Fig. 116 is called the auxiliary circle for the point P and will be used to explain the general terms employed in the analysis of Simple Harmonic Motion.

It will be appreciated that whilst P^1 makes one complete revolution the point P will make one *complete oscillation* across the diameter AB, i.e. from A to B and back again to A. Hence the time for one revolution of P^1 is equal to the time for one complete oscillation or period of P, and this is called the *Periodic Time T.*

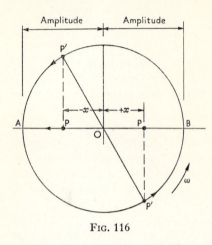

Amplitude   Amplitude

FIG. 116

Now distance moved by P^1 in one revolution $= 2\pi$ radians.

Then, since $\omega = \dfrac{2\pi}{T}$,

$$T = \frac{2\pi}{\omega} \text{ seconds.}$$

The reciprocal of T—the time for one oscillation —will give the *frequency* or number of oscillations per second.

Therefore frequency $n = \dfrac{1}{T} = \dfrac{\omega}{2\pi}$ oscillations or cycles per second.

The *amplitude* of oscillation of P is the *radius* of the auxiliary circle and is simply the maximum displacement of P from the centre of oscillation O.

From Chapter III we know that *throughout its travel* P^1 will be subjected to a constant *centripetal* acceleration of $\omega^2 r$ directed towards O. Then, as shown in Fig. 117—at the instant when P is at a distance $+x$ from O the component acceleration of P^1 *along AB* will be *directed towards O* (i.e. in the opposite direction to that in which x increases). Allowing for this difference in direction between the acceleration and the displacement by means of a minus sign we have :

Component acceleration of $P^1 = -\omega^2 r \cos \theta$.

This, therefore, will be the actual acceleration of the point P. Clearly, from Fig. 117, $r \cos \theta = x$, and therefore, the

$$\textbf{Acc. of P} = -\omega^2\textbf{\textit{x}}. \qquad . \qquad . \qquad (1)$$

By equation (1) we have thus established the two requirements for S.H.M., viz.—the acceleration of P is directed towards

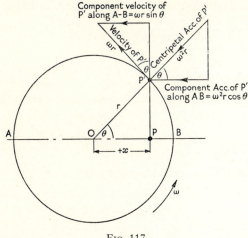

Component velocity of P' along A-B = $\omega r \sin \theta$

Component Acc.of P' along A B = $\omega^2 r \cos \theta$

FIG. 117

the fixed point O and is proportional to x - the distance from the fixed point.

That these conditions are satisfied for every position of P^1 is evident from the auxiliary circle and the fact that the acceleration of P^1 is always directed radially inwards. Expressed in words we have :

$$(\text{Acceleration of P}) = - (\omega^2 \times \text{Displacement of P}),$$

and hence

$$\omega^2 = - \left(\frac{\text{Acceleration of P}}{\text{Displacement of P}} \right) = \frac{\text{a Constant since}}{\omega \text{ is constant.}}$$

Now from the periodic Time equation

$$T = \frac{2\pi}{\omega}$$

and on substituting for ω we obtain :

$$T = 2\pi \sqrt{\frac{\text{Displacement}}{\text{Acceleration}}}$$

and frequency

$$n = \frac{1}{2\pi} \sqrt{\frac{\text{Acceleration}}{\text{Displacement}}}.$$

Since P has a varying acceleration its velocity will also vary. The velocity of P^1 at right angles to $OP^1 = \omega r$ and the horizontal component of this velocity, viz. $\omega r \sin \theta$, will be the actual velocity of P at any given instant, see Fig. 117.

But from the right-angled triangle OP¹P,

$$\sin \theta = \frac{\sqrt{r^2 - x^2}}{r}.$$

Therefore velocity of P for any displacement x from the centre of oscillation is given by

$$\text{Velocity} = \omega r \frac{\sqrt{r^2 - x^2}}{r}$$

or   **Velocity directed towards O** $= \omega \sqrt{r^2 - x^2}.$.   (2)

Equations (1) and (2), therefore, given the acceleration and velocity of the point P which moves about point O with simple harmonic motion.

Acc. of P $= -\omega^2 x = \omega^2 r \cos \theta$ *Directed towards O*
Velocity of P $= \omega r \sin \theta$ *Directed towards O.*

Since both ω and r are constant these equations show that the variation of acceleration with respect to time can be represented by a cosine curve and the variation of velocity by a sine curve. The auxiliary circle clearly indicates that point P passes through O with maximum velocity and zero acceleration ($\theta = 90°$). It then approaches point A with reducing velocity, and at A it has zero velocity and maximum acceleration directed towards the centre of oscillation.

Most free vibrations found in engineering have the characteristics just described so that the first step in any problem is to investigate the relationship which exists between Acceleration and Displacement. If this relationship is constant for all points in the motion, we can then infer that S.H.M. conditions exist.

This approach is illustrated in the examples which follow.

32. Load on a Vertical Spring.—Consider a Vertical Spring carrying a load W, as shown in Fig. 118.

It will be assumed that the spring is perfectly elastic and obeys Hooke's Law, and also that the mass of the spring is very small compared with that of the load.

Let the spring have a stiffness k,

i.e. $k = $ The Load required to produce unit extension of the spring.

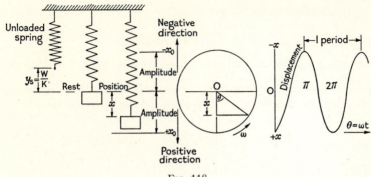

Fig. 118

When the load W is first gradually attached to the spring it will cause it to deflect an amount given by $\frac{W}{k}$, and this is called the static deflection of the loaded spring y_s.

Hence

$$y_s = \frac{W}{k}.$$

In its rest position the load W is balanced by the tension in the spring.

If, now, the load is pulled further downwards a distance x_o and then released, it will oscillate between the limits of $+ x_o$ and $- x_o$.

At the rest position the tension in the spring is equal to W.

At any postition x below the rest position the tension in the spring is greater than the weight W by an amount kx.

When the load is released this excess tension kx will accelerate the load—bringing it first to rest at $- x_o$ and then reversing its direction so that the load once again passes through the rest position.

Now by Newton's Second Law of Motion we have :

$$\text{Force} = \text{Mass} \times \text{Acceleration}$$

$$\therefore \quad - kx = \frac{W}{g} f$$

Mass in slugs if k is in lb. wt./ft.

The minus sign indicates that the load moves in the direction of x decreasing.

By rearrangement we get

$$\frac{f}{x} = -\frac{kg}{W} = \text{a Constant},$$

since k, g and W are all constants for the given load and spring,

and comparing this with the previous article it is thus established that the vibrating weight moves with simple harmonic motion.

$$\therefore \frac{kg}{W} = \omega^2$$

and
$$T = \frac{2\pi}{\omega} = 2\pi\sqrt{\frac{W}{kg}}.$$

But from
$$\frac{W}{k} = y_s$$

$$T = 2\pi\sqrt{\frac{y_s}{g}}.$$

It is important to note that the periodic time and hence the frequency depend *only* on the value of the static deflection which is itself dependent only on the values of the spring stiffness and load W.

Therefore, since the periodic time is independent of the amplitude of the motion, the actual amount of the initial displacement of the spring is of no consequence—providing Hooke's Law still holds.

Now, although the amplitude of the oscillation diminishes with time, the periodic time remains unaltered and explains why simple harmonic motion is also termed isochronous—i.e. repeated motion in the same time interval.

Example 44

A helical spring of negligible weight is required to support a load of 150 lb. If the spring requires a load of 3·5 lb. in order to give an extension of 0·01 inch, what will be the frequency of natural vibration when the 150-lb. weight is displaced 0·5 inch and then released ? Also determine the velocity of the weight when it is 0·25 inch below its rest position.

Solution

$$\text{Stiffness of spring } k = \frac{3\cdot5}{0\cdot01} = 350 \text{ lb./inch.}$$

$$y_s = \frac{W}{k} = \frac{150}{350} = 0.428 \text{ inch.}$$

Then
$$T = 2\pi\sqrt{\frac{y_s}{g}} = 2\pi\sqrt{\frac{0.428}{32.2 \times 12}} = 2\pi \times 0.033,$$

$$\therefore \ T = 0.21 \text{ second.}$$

Frequency $n = \dfrac{1}{T}$,

$$\therefore \textbf{ Frequency} = \textbf{4.77 osc./sec. or 286 osc./min.}$$

To determine the velocity of the weight when it is 0.25 inch below its rest position—as shown diagrammatically in Fig. 119—we have velocity

$$v = \omega\sqrt{r^2 - x^2}.$$

Since
$$T = \frac{2\pi}{\omega} = 2\pi \times 0.033 \text{ from above}$$

we have
$$\frac{1}{\omega} = 0.033 \text{ and } \omega = \frac{1}{0.033} = 30 \text{ rad./sec.}$$

$$v = 30\sqrt{\left(\frac{0.5}{12}\right)^2 - \left(\frac{0.25}{12}\right)^2},$$

r and x in ft. when v in ft./sec.

$$\therefore \textbf{ v} = \textbf{1.082 ft./sec.}$$

When the load on any spring is caused to oscillate it is evident that the spring must also oscillate with it. The effect of the spring on the frequency of oscillation need not be considered when the weight of the spring is negligible—as in this problem. When this is not the case—as for a heavy spring—it can be shown that the effect of the spring can be allowed for by adding one-third of its weight to the load W, and then using the relationships for a light spring as given above.

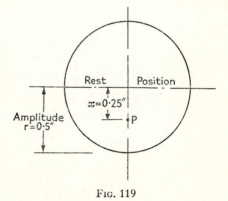

Fig. 119

Example 45

A load is supported by two springs and is constrained to oscillate freely in a vertical direction. When the springs are connected in parallel the frequency of oscillation is found to be 123 osc. per min. and when they are connected in series 60 osc./min. If the stiffness of one spring is 10 lb. per inch what is the stiffness of the other ?

Solution

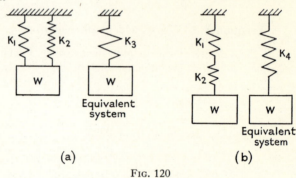

FIG. 120

Let the spring stiffnesses be k_1 and k_2.

Parallel Case

When the load W is displaced, clearly each spring will be deflected the same amount—Fig. 120 (*a*).

Hence for a displacement x :

Spring Force acting on load W $= -(k_1 + k_2)x$.

Then from theory Spring Force $=$ Mass $\times$ Acc.

$$\therefore \frac{\text{Acc.}}{\text{Displacement } x} = -\frac{(k_1 + k_2)}{\text{W}} g = \text{a constant.}$$

This compares with $-\dfrac{k_3}{\text{W}}g$ for a single spring.

Therefore $k_3 = k_1 + k_2 =$ stiffness of the equivalent spring giving the same frequency of oscillation.

Then $$\text{T} = 2\pi \sqrt{\frac{\text{W}}{k_3 g}} = 2\pi \sqrt{\frac{\text{W}}{(k_1 + k_2)g}},$$

and $$n = \frac{1}{2\pi} \sqrt{\frac{(k_1 + k_2)}{\text{W}} g}. \qquad . \qquad . \qquad . \qquad . \qquad (1)$$

Series Case

For the single equivalent spring $y_s = \dfrac{W}{k_4}$ —Fig. 120 (b).

Then since each spring carries the load W, the total static deflection will equal the sum of the separate static deflections of each spring,

$$\text{i.e. } y_s = \frac{W}{k_4} = \frac{W}{k_1} + \frac{W}{k_2}.$$

Therefore, for springs in series :

$$\frac{1}{k_4} = \frac{1}{k_1} + \frac{1}{k_2} = \frac{k_1 + k_2}{k_1 k_2}.$$

Then $\qquad\qquad T = 2\pi\sqrt{\dfrac{W}{k_4 g}} = 2\pi\sqrt{\dfrac{W(k_1 + k_2)}{g}\,\dfrac{}{k_1 k_2}},$

and $\qquad\qquad n = \dfrac{1}{2\pi}\sqrt{\dfrac{g}{W}\left(\dfrac{k_1 k_2}{k_1 + k_2}\right)}.$ $\qquad$. $\qquad$. $\qquad$. $\qquad$. (2)

Hence ratio of frequencies is given by equation (1) divided by equation (2).

$$\therefore \frac{n \text{ (parallel)}}{n \text{ (series)}} = \frac{\dfrac{1}{2\pi}\sqrt{\dfrac{(k_1 + k_2)}{W}g}}{\dfrac{1}{2\pi}\sqrt{\left(\dfrac{k_1 k_2}{k_1 + k_2}\right)\dfrac{g}{W}}}$$

or $\qquad\qquad \left(\dfrac{n \text{ parallel}}{n \text{ series}}\right)^2 = \dfrac{(k_1 + k_2)^2}{k_1 k_2}$

$$\text{i.e. } \left(\frac{123}{60}\right)^2 = \frac{k_1{}^2 + k_2{}^2 + 2k_1 k_2}{k_1 k_2}$$

$$\therefore \; k_1{}^2 + k_2{}^2 - 2\cdot2\,k_1 k_2 = 0.$$

Putting $\qquad k_1 = 10$ lb./inch, gives $100 + k_2{}^2 - 22k_2 = 0.$

Solving the quadratic for k_2 gives

$$\mathbf{K_2 = 15\cdot6 \text{ lb./inch.}}$$

Example 46

The static deflection at the end of a cantilever when supporting an end load is given by $WL^3/3EI$, where W is the load, L the length of the cantilever and EI is a constant for the cantilever called the flexural rigidity. If the end load is displaced from its equilibrium position and then released, show that it will perform simple harmonic motion. Hence calculate for a cantilever which gives a static deflection of $1\frac{1}{4}$ inches for an end load of 5 lb. in a span of 20 inches, the length necessary to give a periodic time of 1 second with the 5 lb. on the end.

Solution

Let k be the beam stiffness, i.e. the end load necessary to give unit deflection of the end of the cantilever.

Then if the end load is displaced a distance x from its equilibrium position y_s, the restoring force acting on the load $= -kx$.

$$\therefore \quad -kx = \frac{W}{g} f.$$

or $\quad \dfrac{\text{Displacement}}{\text{Acceleration}} = \dfrac{x}{f} = -\dfrac{W}{kg} = \text{a constant.}$

Hence the cantilever will act in the same manner as a spring if the end load is given an initial displacement and then released and allowed to oscillate.

In this case $\quad \dfrac{W}{k} = y_s = \dfrac{WL^3}{3EI} = CL^3$

Since $\dfrac{W}{3EI}$ is a constant in this example.

$$\therefore \quad T = 2\pi \sqrt{\frac{CL^3}{g}}.$$

For the 20-inch span deflection is $1\frac{1}{4}$ inches,

$$\therefore \quad 1\tfrac{1}{4} \text{ in.} = C20^3$$

or $\quad C = \dfrac{5}{4.20^3} = 1 \cdot 56 \times 10^{-4}.$

Then for $T = 1$ second, we have from the equation for T,

$$1 = 2\pi \sqrt{1 \cdot 56 \times 10^{-4} \frac{L^3}{12 \cdot g}},$$

g in in./sec.^{2} for L in inches.

$$\therefore \quad L = 39 \cdot 6 \text{ inches.}$$

Example 47

A glass U-tube is mounted with its limbs vertical and contains a liquid. A slight pressure is applied to one of the limbs and then released, causing the column of liquid in the tube to oscillate.

Show that the liquid moves with simple harmonic motion and that the period is independent of the density of the liquid and of the bore of the tube, but is a function of the overall length of the liquid column.

Hence calculate the period of oscillation of a column of mercury of 36 inches over-all length.   Take $g = 32.0$ ft./sec.²   (U.E.I.)

Solution

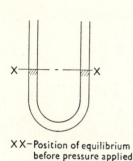

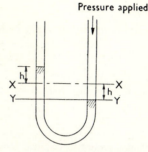

Pressure applied

X X—Position of equilibrium
before pressure applied
(i)

Y Y—Position of equilibrium
with pressure applied
(ii)

Fig. 121

Let l = total length of liquid column ; d = dia. of tube ; ρ = mass density of liquid ; ω = specific weight of liquid.

Fig. 121 (i) shows the liquid column at rest—the common liquid surface being represented by the line XX.

If the liquid in the right-hand limb is now depressed a distance h *below* XX, then, since the tube is uniform, the liquid in the left-hand limb will rise h *above* XX, see Fig. 121 (ii).

Hence the liquid displacement from XX is given by h for each limb.

Consider now the line YY through the surface of the liquid in the right-hand limb.   Whilst the slight pressure is acting, equilibrium will exist across YY—the applied force just balancing the force due to the weight of liquid length $2h$ above YY in the left limb.

When the pressure is removed the force due to the now unbalanced weight of liquid in the left limb will cause the *whole* liquid column l to move.

Then force causing motion   = weight of the liquid column of length $2h$. But weight of liquid column = Volume × Specific Weight.

$$\therefore \text{ Force} = \left(\frac{\pi d^2}{4} \cdot 2h\right)\omega.$$

Since $\qquad\qquad\qquad \omega = \rho g$ (see Chapter X)

we have $\qquad\qquad$ Force $= \dfrac{\pi d^2}{4} \cdot 2h \cdot \rho g.$

Also mass of liquid which is moved by this force

$$= \dfrac{\pi d^2}{4} l \rho.$$

Then applying the equation Force = Mass × Acc. we have

$$\dfrac{\pi d^2}{4} \cdot 2h\rho g = \dfrac{\pi d^2}{4} l \rho \times \text{Acc.}$$

Hence by rearrangement :

$$\text{Acc.} = \dfrac{2h}{l} g.$$

But displacement of column from equilibrium position $= h$,

$$\therefore \; \dfrac{\text{Displacement}}{\text{Acceleration}} = \dfrac{h}{\dfrac{2h}{l} g} = \dfrac{l}{2g}, \text{ which is a constant.}$$

From article 31, since the relationship between displacement and acceleration is constant, the liquid column moves with S.H.M. of period

$$T = 2\pi \sqrt{\dfrac{l}{2g}}.$$

This clearly shows that the periodic time is independent of the density ρ and the bore d and depends only on the length of the liquid column l.

For a column of length 36 inches :

$$T = 2\pi \sqrt{\dfrac{36}{12 \times 2 \times 32}} = \dfrac{\pi \sqrt{3}}{4},$$

$$\therefore \textbf{ Periodic Time} = \mathbf{1 \cdot 36} \textbf{ seconds.}$$

33. The Simple Pendulum.—A simple pendulum consists of a heavy particle or bob of negligible dimensions at one end of a weightless cord, the other end of which is fixed. The particle oscillates in a vertical plane under the action of gravity. Clearly, such an ideal pendulum cannot exist practically, but its motion is often referred to when considering the motion of actual pendulums (see next article).

For the simple pendulum in Fig. 122 the length of cord is l and at the given instant it is inclined at an angle θ to the vertical. Whilst the particle moves in a circular path of radius l the only forces acting upon it are its weight W vertically downwards and the tension in the cord T acting at right angles to the tangent.

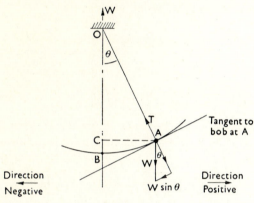

FIG. 122

Since the force T has no component along the tangent the only force acting on the particle in this direction is the component of the weight of the bob, i.e. W sin θ.

Hence the linear acceleration of the particle along the tangent is given from

$$- W \sin \theta = \frac{W}{g} \text{ Acc.}$$

or $\qquad$ Acc. along the tangent $= - g \sin \theta$.

Now if θ is small $(\theta \ngtr 10°)$ sin $\theta \simeq \theta$ and the acceleration is approximately $g\theta$.

Then since $\qquad l\theta = \text{Arc AB}$

$$\theta = \frac{\text{Arc AB}}{l}.$$

We thus have *Acc.* along tangent is

$$\simeq - g \cdot \frac{\text{Arc AB}}{l}.$$

A.M.—L

But the arc AB gives the displacement of the particle from the fixed point B.

$$\therefore \frac{\text{Displacement}}{\text{Acceleration}} \simeq -\frac{l}{g} \simeq \text{a Constant.}$$

Therefore the motion of the simple pendulum is approximately simple harmonic and the period is given by

$$T = 2\pi\sqrt{\frac{l}{g}},$$

and the frequency by

$$n = \frac{1}{2\pi}\sqrt{\frac{g}{l}}.$$

Alternatively, using the notation of article 27, chapt. IV ; the torque tending to restore the particle to its equilibrium position at B is given by :

$$\text{Torque} = -W. \times AC = -Wl \sin \theta,$$

which, for small values of θ, becomes

$$\text{Torque} \simeq -Wl\theta. \qquad . \qquad . \qquad . \qquad (1)$$

Also Torque $= I_o\alpha$ where I_o is the mass moment of inertia of the particle about O and α the angular acceleration of the cord.

Since $I_o = ml^2 = \dfrac{W}{g}l^2$, where m is the mass of the particle in slugs.

$$\text{Torque} = \frac{W}{g}l^2\alpha. \qquad . \qquad . \qquad . \qquad (2)$$

Equating (1) and (2) gives

$$\frac{\theta}{\alpha} = \frac{\text{Angular Displacement}}{\text{Angular Acceleration}} \simeq -\frac{l}{g} \simeq \text{a Constant.}$$

Hence, as before,

$$T = 2\pi\sqrt{\frac{l}{g}}.$$

34. The Compound Pendulum.—Any body suspended vertically and capable of free rotation under the action of gravity is termed a compound pendulum.

Consider the body in Fig. 123.

Let O be the 'centre of suspension', W the weight of the body, M, its mass and K_o the radius of gyration about an axis through O perpendicular to the plane of motion.

Let 'a' be the distance from O to the centre of gravity G.

Then, from the previous article, when the body is displaced an angle θ from its equilibrium position :

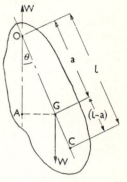

Restoring Torque $T = - W . AG$

$$= - Wa \sin \theta \simeq - Wa\theta.$$

Also $\quad T = I_o\alpha = MK_o{}^2\alpha = \dfrac{W}{g}K_o{}^2\alpha.$

$$\therefore \frac{W}{g}K_o{}^2\alpha \simeq Wa\theta.$$

FIG. 123

Hence $\qquad \dfrac{\theta}{\alpha} = \dfrac{\text{Angular Displacement}}{\text{Angular Acceleration}} \simeq \dfrac{K_o{}^2}{ga},$

$\simeq$ a constant, since K_o and ga are all constants.

Therefore the motion of the compound pendulum is approximately simple harmonic having period

$$T = 2\pi\sqrt{\frac{K_o{}^2}{ga}} \text{ and frequency } \frac{1}{2\pi}\sqrt{\frac{ga}{K_o{}^2}}.$$

If K_G be the radius of gyration of the body about an axis through the centre of gravity perpendicular to the plane of motion, then, by the Parallel Axes Theorem,

$$I_o = I_G + Ma^2$$
or $\qquad\qquad MK_o{}^2 = MK_G{}^2 + Ma^2$
and $\qquad\qquad K_o{}^2 = K_G{}^2 + a^2.$

Substituting for $K_o{}^2$ in the above equations gives

$$T = 2\pi\sqrt{\frac{K_G{}^2 + a^2}{ga}},$$

and $\qquad\qquad n = \dfrac{1}{2\pi}\sqrt{\dfrac{ga}{K_G{}^2 + a^2}}.$

If we compare these equations with those for a simple pendulum we have

$$T = 2\pi\sqrt{\frac{K_G^2 + a^2}{ga}} \text{ and } T = 2\pi\sqrt{\frac{l}{g}}.$$

Then, clearly, the equivalent length of simple pendulum which would have the same periodic time and frequency as the compound pendulum is given by :

$$l = \frac{K_G^2 + a^2}{a}.$$

Hence, if a point C is taken in Fig. 123 such that the length OC is equal to that of a simple pendulum which has the same periodic time, then

$$OC = l = \frac{K_G^2 + a^2}{a}. \qquad . \qquad . \qquad . \qquad (1)$$

The point C is called the 'centre of oscillation'.

Suppose now that the centre of suspension O and the centre of oscillation are now reversed and the pendulum is allowed to oscillate about an axis through C parallel to the original axis through O. Then the length CG corresponds to 'a' in the original equations—see Fig. 123,

i.e. $CG = (l - a)$ must be substituted for 'a',

$$\therefore \ T = 2\pi\sqrt{\frac{(l - a)^2 + K_G^2}{g(l - a)}}.$$

From (1) $(l - a) = \dfrac{K_G^2}{a}.$

Hence on substitution

$$T = 2\pi\sqrt{\frac{(K_G^2/a)^2 + K_G^2}{g\dfrac{K_G^2}{a}}},$$

and $$T = 2\pi\sqrt{\frac{K_G^2 + a^2}{ga}},$$

which is identical with that for the time of swing about the axis through O.

Therefore it has been shown that the centres of oscillation and suspension are interchangeable.

Example 48

Show that the periodic time for a compound pendulum is a minimum when the centre of suspension is at a distance equal to K_G from the centre of gravity.

A connecting rod AB, when suspended from a horizontal knife edge at A and allowed to oscillate, performs 60 oscillations per minute. The centre of gravity of the rod is 6 inches from the knife edge at A. Determine the radius of gyration of the rod about an axis through the C.G. and parallel to the knife edge. Also calculate the frequency of small angular oscillations when the rod is suspended at B.

Solution

In the equation $$T = 2\pi\sqrt{\dfrac{K_G^2 + a^2}{ga}}$$

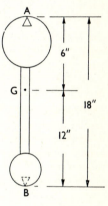

'a' represents the distance of the C.G. from the centre of suspension.

If T is to be a minimum then $\dfrac{K_G^2 + a^2}{a}$ must

also be a minimum (since g is a constant).

Then by differentiation of $\dfrac{K_G^2 + a^2}{a}$,

$$\frac{dT}{da} = -\frac{K_G^2}{a^2} + 1 = \text{zero for a minimum,}$$

or $$K_G = a.$$

Fig. 124

Hence the periodic time is a minimum when the centre of suspension is at a distance equal to K_G from the centre of gravity.

From the equation frequency $= \dfrac{1}{2\pi}\sqrt{\dfrac{ga}{K_G^2 + a^2}}$,

'a' $= 6$ inches when suspended from A, and the number of oscillations/sec. is 1,

$$\therefore 1 = \frac{1}{2\pi}\sqrt{\frac{(32\cdot2\times12)6}{K_G^2 + (6)^2}}.$$

Hence $$K_G^2 = 22\cdot7 \text{ and } K_G = 4\cdot77 \text{ inches.}$$

When suspended from B, 'a' $=12$ inches (Fig. 124),

$$\therefore \; n_B = \frac{1}{2\pi}\sqrt{\frac{(32 \cdot 2 \times 12)12}{22 \cdot 7 + 12^2}}$$

$$= 0 \cdot 841 \text{ osc./sec.}$$

$$\therefore \; n_B = \mathbf{50 \cdot 5 \text{ osc./min.}}$$

EXERCISES V

1. Define Simple Harmonic Motion.

A vertical helical spring having a stiffness of 4 lb./in. is clamped at its upper end and carries a weight of 6 lb. attached to the lower end. The weight is displaced vertically a distance of 1 inch from the equilibrium position and released. Determine the period of the resultant vibration and the maximum velocity attained by the weight.   U.E.I.

2. Derive an expression for the periodic time of a body moving in a straight line with simple harmonic motion. The stiffness of a helical spring is 150 lb. per foot of extension. Find its periodic time of vibration when supporting a weight of 10 lb.

If the weight is initially displaced $\frac{1}{2}$ inch, what would be the maximum velocity of the weight during its motion ?   U.E.I.

3. Obtain the relationship between the linear acceleration and displacement of a body executing simple harmonic motion and deduce a formula for the frequency of vertical oscillations of a close coiled helical spring when it is fixed at its upper end and is carrying a load at the lower end.

A vertical spring of stiffness 16 lb. per inch is fixed at its upper end and carries at its lower end a 48-lb. weight. This weight is pulled down and then released, setting up vertical oscillations. Determine (a) the natural frequency of vertical oscillations, (b) the velocity of the 48-lb. weight when it is 0·375 inch below the equilibrium position, given that the amplitude of oscillation is 1·5 inch.

Neglect the weight of the spring and air resistance.

4. Show that for a weight of W lb. attached to a spring of stiffness 'S', the periodic time of small oscillations is given by $t = 2\pi\sqrt{\dfrac{W}{Sg}}.$

A weight of 5 lb. is attached to a spring of stiffness 15 lb. per foot. The spring is extended 3 inches beyond the equilibrium position and the weight then released. Determine :

(a) the frequency of oscillations per minute ;

(b) the velocity of the weight when it is 1 inch from its equilibrium position.

5. A cantilever spring of negligible weight is 12 inches long and carries a load of 5 lb. at the free end. When an additional load of 20 lb. is applied to the free end the cantilever is deflected downwards 1 inch from the equilibrium position. If this additional load is suddenly removed, determine the period of vibration of the 5-lb. load and also its maximum velocity.

6. Define linear simple harmonic motion and find an expression for the velocity V when the displacement from the mid-position is X.

The ram of a pump may be taken to move with linear simple harmonic motion. Find the amplitude or half travel if, for velocities of 8 ft./sec. and 7 ft./sec., the displacements from the mid-position are 2·0 and 2·5 inches respectively. I.Prod.E.

7. A piston moving with simple harmonic motion passes through two points A and B, 16 inches apart, with the same velocity, having taken 2 seconds in passing from A to B. Three seconds later it returns to point B. Determine (a) the period and amplitude of the oscillation, (b) the maximum acceleration of the piston.

8. A body obtains simple harmonic motion of amplitude 'a' from a crank rotating at N r.p.m. Show how to express the displacement, velocity and acceleration of the body in terms of these quantities. The crankshaft of a steam engine rotates at 250 r.p.m. The eccentric on the shaft moves the slide valve with simple harmonic motion over a stroke length of 6 inches. The valve opens to steam when it has moved $1\frac{1}{4}$ inches past its mid-position, and opens to exhaust on the return stroke when $\frac{3}{4}$ inch beyond the mid-position. The valve weighs 50 lb. and the frictional force acting against the motion may be taken as constant at 150 lb. Determine the angle turned through by the eccentric between the openings to steam and exhaust, and the driving force required when opening to steam occurs. I.Mech.E.

9. A body weighing 80 lb. moves with simple harmonic motion and makes 60 strokes per minute. The length of stroke is 18 inches. Determine maximum accelerating force acting on the body and draw a diagram to show the accelerating force for each position in the line of stroke.

10. Find an expression for the acceleration, at displacement X from its mid-position, of a body moving in a straight line with 'Simple Harmonic Motion'.

The piston of an engine has a stroke of 3 feet and the reciprocating parts weigh 550 lb. If the crank shaft makes 250 r.p.m., find the force to overcome the inertia of the reciprocating parts at the ends of the stroke, and also the inertia force and the velocity of the reciprocating parts when the crank has turned 30 degrees from either dead centre. U.E.I.

11. Part of a machine has a reciprocating motion which is simple harmonic making 200 complete oscillations in one minute. It weighs 10 lb. and has a stroke of 9 inches. Find (a) the accelerating force upon it in pounds and its velocity in ft./sec. when it is 3 inches from mid-stroke, (b) the maximum accelerating force and (c) the maximum velocity.

12. A body weighing 50 lb. moves with simple harmonic motion with a frequency of 100 complete oscillations per minute. The maximum velocity attained by the body is 15 ft./s. Determine (a) the amplitude of the motion, (b) the displacement of the body from the mid-position when the force acting on the body is 150 lb. weight, (c) the time taken by the body to move from the extremity of the motion to a point mid-way between the extreme and mid-position of the motion.   U.L.C.I.

13. Define 'simple harmonic motion'.

A body weighs 100 lb. and moves along a straight line with simple harmonic motion. At distances of 1 foot and 2 feet from the mid-point of the oscillation, the velocities of the body are 3 ft./s. and 2 ft./s. respectively.

Determine (a) the amplitude of the motion, (b) the time for one complete oscillation of the body, (c) the force acting on the body when it is 2 feet from the extremity of the oscillation.   U.L.C.I.

14. Show that the period of vibration T of a simple pendulum of length L, is given by $T = 2\pi \sqrt{\dfrac{L}{g}}$.

Hence or otherwise deduce the length in inches of the simple pendulum having a period of two seconds.   U.E.I.

15. Derive from first principles an expression for the periodic time of a compound pendulum in terms of its Radius of Gyration K_G, with respect to the centre of gravity. Hence show that this time is a minimum when the point of suspension is at a distance equal to K_G from the centre of gravity.

16. Prove from first principles that the radius of gyration of a uniform rod of length L, about an axis through the centre of gravity and perpendicular to its length, is $\dfrac{L}{2\sqrt{3}}$. A uniform metal rod is 6 feet long and is oscillated as a compound pendulum about axes perpendicular to its length. Calculate :

    (a) the periodic time of swing when the axis is 2 feet from one end ;
    (b) the position of the axis for the periodic time to be a minimum ;
    (c) the minimum periodic time.

17. A small flywheel of weight 200 lb. was suspended in a vertical plane as a compound pendulum. When the distance to the centre of gravity from the knife edge support was 10 inches, the flywheel made 100 oscillations in 127 seconds. Working from first principles obtain a value for the moment of inertia of the flywheel about an axis through the centre of gravity.

18. A uniform plate, 1 foot square, is pivoted vertically about one corner and caused to oscillate. Determine the frequency of small angular oscillations.

19. A link in a mechanism has a weight of 120 lb. and swings about a centre 'A' which is 3 feet from the centre of gravity of the link. When the link was oscillated as a compound pendulum about 'A', 10 complete swings were noted to take 27·8 seconds. If the link is speeded from 50 rev./min. to 76 rev./min. in 2 seconds, calculate the angle turned through and the torque which is necessary to achieve this.

20. A connecting rod of weight 80 lb. and 32 inches long between centres is suspended vertically. The time for 60 oscillations is found to be 92·5 seconds when the axis of oscillation coincides with the small end centre and 88·4 seconds when it coincides with the big end centre. Find :

  (*a*) the moment of inertia of the rod about an axis through the centre of gravity ;

  (*b*) the distance of the centre of gravity from the small end centre.

Chapter VI

STRESS AND STRAIN

So far our discussion has been mainly concerned with the relationship between the forces applied to a body and the resulting motion that takes place. Our next step must be to inquire into the distorting effects which such forces produce on the bodies themselves. The extent of this inquiry is suggested by our knowledge that the effects produced by any force will depend not only on the way in which the force is applied but also on the material of which the body is made.

35. Direct Stress.—If an external force is applied to a body, as in Fig. 125, the body will be distorted and is said to be strained. The body will remain in equilibrium under the action of this force only so long as the *internal* forces can resist the deformation which accompanies the application of that force.

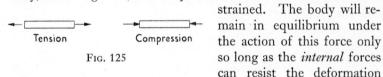

Tension   Compression

Fig. 125

A knowledge of the magnitude of the internal forces which the material of a given member can offer is, therefore, necessary, if we are to ensure that equilibrium is maintained for all conditions of loading.

Consider a bar—Fig. 126—subjected to a tensile load P. If we imagine the bar to be divided into two parts across a section XX at right angles to its axis, we can examine the left-hand section by itself. It is evident that for equilibrium to exist the external force P acting to the left must be balanced by an equal force acting to the right—Fig. 126. This force to the right is supplied by the individual molecular forces representing the action of the particles of the right-hand section on the particles of the left-hand section. A similar argument will hold for the right-hand end. Therefore, it is evident that between the molecules of the bar on either side of any transverse section there exists a

mutual action consisting of *equal* and *opposite* internal forces—
each balancing the external force P.

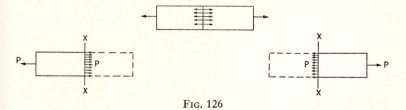

FIG. 126

The internal force transmitted per unit of cross-sectional area
is called the intensity of stress—or, more commonly, just the
stress—so that for any cross-section such as XX, having an area A,

$$\text{Stress } f = \frac{\text{Load}}{\text{Area}} = \frac{P}{A}.$$

The stress in Fig. 126 is shown uniformly distributed over the
section XX, and this may always be taken to be the case providing
the section chosen is some distance from the point of application
of the load—see Fig. 127.

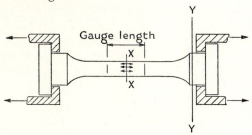

Stress distribution across XX uniform
Stress distribution across YY non-uniform

FIG. 127

Tensile and compressive stresses acting normal to given
sections are usually referred to as *normal* or *direct* stresses.

Units of stress lb./in.^{2} ; tons/in.^{2} ; gm./cm.^{2}.

36. Shear Stress.—Since we are not only concerned with
problems of direct stress it will be instructive if we now investigate
the stress conditions on a section XX within the material which is

inclined to the axis of the bar as in Fig. 128. Again let us consider a tensile loading and choose an oblique section XX at a sufficient distance from the end of the bar to give a uniform stress distribution across the section.

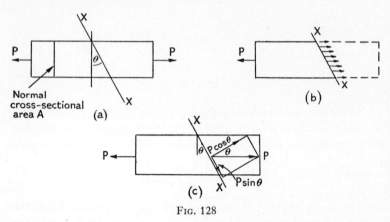

Fig. 128

If A is the cross-sectional area of the bar *normal* to its axis and if θ is the inclination of the oblique plane to the normal section, then clearly the area across XX will equal $A/\cos\theta$ (i.e. it is greater than A).

The load transmitted by the oblique section (which for equilibrium equals the horizontal external force P) can be resolved into components normal and tangential to XX—Fig. 128 (c). These components of P provide a normal force $P\cos\theta$ and a tangential force $P\sin\theta$, both acting on the area $A/\cos\theta$.

Hence, from our definition of stress we have a normal tensile stress—perpendicular to XX—of magnitude :

$$\frac{\text{Load}}{\text{Area}} = \frac{P\cos\theta}{A/\cos\theta} = \frac{P}{A}\cos^2\theta,$$

and a tangential stress of magnitude :

$$\frac{\text{Load}}{\text{Area}} = \frac{P\sin\theta}{A/\cos\theta} = \frac{P}{A}\sin\theta\cos\theta.$$

Since $\sin\theta\cos\theta$ $\qquad = \dfrac{\sin 2\theta}{2},$

we get tangential stress $\qquad = \dfrac{P}{A}\dfrac{\sin 2\theta}{2},$

thus—Normal stress on $XX = \dfrac{P}{A} \cos^2 \theta,$

and Tangential Stress on $XX = \dfrac{P}{A} \dfrac{\sin 2\theta}{2}.$

These two equations indicate that even for a simple tensile (or compressive) loading it is possible for stresses other than normal stresses to occur within a material.

When θ, the inclination of the plane, is zero these equations show that the normal stress is P/A and the tangential stress is zero, which is the case given in Fig. 126.

When θ is 45° the tangential stress is a maximum, since $\sin 2\theta = \sin 90°$ is a maximum and this gives the maximum value of the tangential stress as half the normal stress value for the case of simple tension, i.e. $\frac{1}{2} \dfrac{P}{A}$ (see article 44).

Further consideration of the tangential stress across XX in Fig. 128 suggests that it has a tendency to 'shear' or slide the right-hand section across the left-hand section of the bar and hence this type of stress is called a *shearing stress*.

Shearing stress can occur on Fig. 128 where it is accompanied by a direct stress on the same plane or it can occur alone as shown in Fig. 129. In this latter case it is referred to as a *pure* shear stress.

Shear stress is force per unit area.

Hence Shear Stress $f_s = \dfrac{F}{A}.$

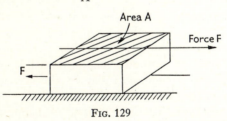

Fig. 129

37. Complementary Shear Stress.—The forces P cos θ and P sin θ in Fig. 128 will introduce normal and tangential stresses on the section XX.

Let the normal stress produced by P cos θ be f_n and P sin θ produce a shearing stress f_s. Then if we imagine a little block of the material *abcd* having one of its faces *cd* coincident with plane XX, we shall have the stress conditions depicted in Fig. 130 (ii).

If this block is to be in equilibrium it will be necessary for the left-hand section of the bar to supply two stresses on the face *ab*. These are :

  (i) a normal stress f_n and
  (ii) a tangential or shear stress f_s.

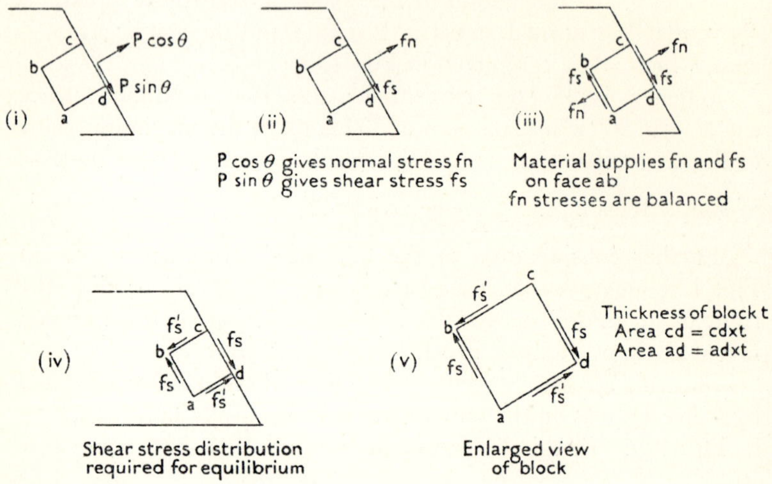

(i)

(ii)

P cos θ gives normal stress fn
P sin θ gives shear stress fs

(iii)

Material supplies fn and fs
on face ab
fn stresses are balanced

(iv)

Shear stress distribution
required for equilibrium

(v)

Enlarged view
of block

Thickness of block t
Area cd = cdxt
Area ad = adxt

FIG. 130

Now whilst a normal stress f_n on face *ab* will completely balance the normal stress f_n on face *cd* it will be evident from the diagram at (iii) that an equal and opposite shear stress on *ab* will introduce a couple tending to turn the block clockwise in the bar. Clearly, this rotation can only be prevented if an equal and opposite couple is brought into play by the action of opposing shear stresses on the faces *bc* and *da*.

Let it be assumed that shear stresses of magnitude $f_s{}^1$ act on these two parallel faces *bc* and *da* (iv).

If the thickness of the bar is t (at right angles to the paper), then the area of face *cd* = $cd \times t$.

Then the shearing force on this area = $f_s \cdot cd \times t$.

This will also be the force on *ab*, and hence these two forces will produce a clockwise couple of magnitude $f_s cd \times t \times ad$.

Similarly the area of face $ad = ad \times t$ and the shearing force on it $= f_s^1 ad \times t$, and this is also the force on da. These two forces produce an anticlockwise couple of magnitude $f_s^1 ad \times t \times cd$.

For equilibrium of the block these two couples must be equal.

$$\text{Therefore, } f_s cd \times t \times ad = f_s^1 ad \times t \times cd,$$

or
$$\mathbf{f_s = f_s^1}.$$

This analysis indicates that whenever shear stresses occur at any plane *within* a material, equal shear stresses are developed on planes at right angles. These stresses are called *complementary shear stresses*.

An appreciation of this very important topic will be helpful when dealing with the theory of torsion in Chapter IX.

38. Tensile and Compressive Strain.—Whenever a member is loaded some deformation is bound to occur resulting in a change in its dimensions. Consider the circular bar in Fig. 131.

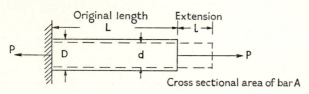

FIG. 131

Under a tensile load an extension in length will take place in the direction of the load which will be accompanied by a decrease in diameter—the reverse occurring in compression.

The proportional change in dimensions is called the *strain*.

$$\text{Hence Longitudinal strain } (e) = \frac{\text{Extension}}{\text{Original Length}} = \frac{l}{L},$$

$$\text{Lateral strain} = \frac{\text{Reduction in dia.}}{\text{Original dia.}} = \frac{D - d}{D}.$$

Since strain is a ratio of dimensions it is clearly *dimensionless*.

39. Shear Strain.—If we consider a uniform shearing stress to act on a rectangular block, as in Fig. 132, it will have the effect of distorting the material to the new position shown dotted. The small angle ϕ, which determines the amount of distortion, is called the *shear strain*.

For small angles $\phi = \dfrac{x}{L}$, ϕ being measured in radians.

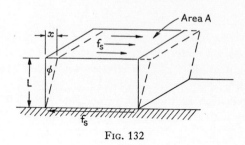

Fig. 132

40. Elasticity and Hooke's Law.—It has been shown that when a body is loaded it suffers distortion. If, when the load is removed, the body returns to its original dimensions the body is said to be *perfectly elastic*.

The property of a body of returning to its original shape when unloaded is called *elasticity*.

A body is only partially elastic if the deformation produced by a load does not all disappear completely when it is unloaded. Such a body is permanently strained and is said to suffer a *permanent set*.

Our simple theories on the behaviour of materials under different loadings are all based on the assumption that the material remains perfectly elastic and obeys Hooke's Law, which states that 'Stress is proportional to Strain'. Experiments on most engineering materials verify the validity of this assumption, within a certain limit, which is different for each material.

Hooke's Law applies equally well for tensile, compressive and shear stresses, and hence we obtain the following relationships—see Figs. 131 and 132:

$$\frac{\text{Tensile stress}}{\text{Tensile strain}} = \frac{f}{e} = \frac{P/A}{l/L} = \frac{PL}{Al} = E \quad \begin{array}{l}\text{(A constant for any}\\ \text{given material).}\end{array}$$

$$\frac{\text{Shear stress}}{\text{Shear strain}} = \frac{fs}{\phi} = \frac{F/A}{x/L} = \frac{FL}{Ax} = G \quad \begin{array}{l}\text{(A constant for any}\\ \text{given material).}\end{array}$$

The constant E, which is the same for both tensile and compressive stresses, is called *Young's Modulus of Elasticity*, and the constant G is called the *Modulus of Rigidity*.

Since strain is a ratio of like dimensions E and G will clearly have the same units as those employed for stress.

The range of stress over which Hooke's Law applies for any material can be obtained by measuring the stress which accompanies increasing values of strain. If the loading is continued beyond this limiting stress up to the point of fracture and a graph of stress against strain is plotted, it will provide valuable information about the mechanical properties of the material as illustrated in the following article.

41. Tensile Test Diagram.—Tensile tests are carried out in testing machines designed to apply truly axial loads. The test specimens are of uniform cross-section and whenever possible conform to the sizes and shapes suggested by the British Standards

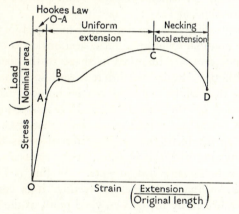

FIG. 133

Institution. The strain is determined over a measured length of specimen called the gauge length which is taken some distance from the end fixings—see Fig. 127. Fig. 133 illustrates a typical stress-strain diagram for ductile materials such as mild steel and wrought iron. (A ductile material is one that can be drawn out without breaking.)

A.M.—M

From O to A the strain is proportional to the stress. The strain in this region is shown greatly magnified in the diagram. 'A' marks the point at which the material deviates from Hooke's Law and hence is called the *limit of proportionality*. (For mild steel this point is found to practically coincide with the elastic limit whilst with other materials—notably the non-ferrous alloys—this is not the case.)

Further loading beyond point A is accompanied by a more rapid increase in strain and the diagram is curved from A to B. At B, a slight drop in load is followed by considerable strain at almost constant load. This action constitutes a yielding of the metal and B is called the *yield point* and the corresponding stress the *yield stress*. Recovery of the material then takes place and the strain continues with increasing load up to C, which is the *maximum load* point. The stress at point C is called the *Ultimate Tensile Stress* or *U.T.S.* Up to this maximum load elongation of the specimen takes place with hardly any visual reduction in cross-section, but from C onwards there is a local reduction in area called 'necking', which is clearly defined as the load falls off to the fracture point at D.

Stress-strain diagrams, as illustrated in Fig. 133, are usually based on the nominal or original cross-sectional area of the specimen and hence they are identical in shape to the load-extension diagrams, although drawn to different scales. However, it is worth noting that whilst the load can be reduced from C to D the stress based on the actual 'necked' cross-sectional area of the specimen would rise considerably above that at C.

In order to establish the points referred to as the limit of proportionality and the elastic limit, it is necessary to employ very sensitive measuring instruments. As this is found to be impractical in ordinary commercial testing, it is convenient to consider the more easily recognised yield point as the limit of the elastic range when estimating the maximum working stress for the material.

42. Working Stress and Factor of Safety.—It should be evident that in order to keep any loaded member free from permanent strain when the load is removed the *working stress* for the material must always be well within the elastic range. From the foregoing discussion this condition is satisfied provided the working stress does not exceed the yield stress and this may be ensured by employing a *factor of safety* in the equation.

$$\text{Working Stress} = \frac{\text{Ultimate Tensile Stress}}{\text{Factor of Safety}}.$$

The value chosen for the factor of safety will depend very much on the conditions of loading and the nature of the material employed. To give some idea of the effect of loading conditions on the working stress we may consider a mild steel having a U.T.S. figure of, say 30 tons/in.² For static or dead load conditions the working stress would probably not exceed 10 tons/in². (F. of S. 3), whereas for suddenly applied or shock loading conditions 2½ tons/in.² would be the maximum permissible working stress figure.

43. Percentage Elongation and Reduction in Area.—The amount of strain which a material can take before fracture occurs gives a measure of its ductility. In a tensile test ductility is defined by the percentage elongation and percentage reduction in area of the fractured specimen.

$$\text{Percentage Elongation} = \frac{\text{Extension of gauge length}}{\text{Original gauge length}} \times 100.$$

Percentage Reduction of area

$$= \frac{\text{Original area} - \text{Final area at fracture}}{\text{Original area}} \times 100.$$

For ductile materials the extension of the specimen is found to be fairly *uniform* throughout the gauge length up until the maximum load is reached. After this load necking of the specimen begins and considerable *local* extension occurs on each side of the fracture.

The total extension will be the sum of both the uniform and the local extensions.

Now the uniform extension, before the maximum load, is proportional to the gauge length, whereas the local extension is a fixed amount—dependent on the cross-sectional area but *independent* of the gauge length. This means that the total extension of a specimen will be more affected by the local extension when the gauge length is *short* than when it is long, and, for this reason, it is always important to state percentage elongation figures in relation to the gauge length employed.

44. The Shape of the Fracture.—The shape of a specimen at fracture can often give useful information regarding the condition of the material. Brittle materials, such as cast iron and brass, show very little extension and fracture with a rather flat, ragged surface whilst ductile materials are characterised by a cup and cone shaped fracture, as illustrated in Fig. 134.

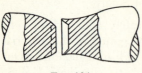

FIG. 134

The shape of this particular fracture indicates that both tensile and shear failure takes place. The inner part of the material fails in tension and the outer part of the cup fails due to the shear stresses being overcome. The angle of the cups is approximately 45°, which supports the theory given in article 36 that maximum shear stresses occur on planes at 45° to the direction of the applied tensile stress.

Example 49

A commercial tensile test on two specimens of the same mild steel, one cold rolled and the other annealed, gave the following results :

	Cold Rolled	Annealed
Yield Load	6·70 tons	3·80 tons
Maximum Load	8·04 tons	5·35 tons
Final length between gauge points	9·375 in.	9·969 in.
Diameter at fracture	0·328 in.	0·273 in.

The original gauge length was 8 inches and the diameter of specimens 0·5 inches.

Determine the yield stress, ultimate tensile stress, percentage elongation and percentage reduction in area for each specimen and tabulate the results to show the affect of the heat treatment on the properties of the material.

Solution

Mild steel, cold rolled.

$$\text{Yield Stress} = \frac{\text{Yield Load}}{\text{Nominal C.S.A.}} = \frac{6·70}{\frac{\pi}{4}(0·5)^2} = 34 \text{ tons/in.}^2$$

$$\text{U.T.S.} = \frac{\text{Max. Load}}{\text{Nominal C.S.A.}} = \frac{8 \cdot 04}{\frac{\pi}{4}(0 \cdot 5)^2} = 41 \text{ tons/in.}^2$$

$$\% \text{ Elongation} = \frac{\text{Increase in G.L.}}{\text{Original G.L.}} = \frac{9 \cdot 375 - 8}{8} = 17 \cdot 2\% \text{ (for 8'' G.L.)}$$

$$\% \text{ Reduction in Area} = \frac{\text{Reduction in Area}}{\text{Original Area}} = \frac{\frac{\pi}{4}(0 \cdot 5^2 - 0 \cdot 328^2)}{\frac{\pi}{4}(0 \cdot 5)^2} = 56 \cdot 9\%.$$

Similar calculations may be made for the annealed specimen and the required table drawn up as follows :

Specimen	Yield Stress tons/in.^{2}	U.T.S. tons/in.^{2}	% Elonga- tion	% Reduction in Area
M.S. Cold Rolled	34·0	41.0	17·2	56·9
M.S. Annealed	19·3	27·3	24·6	70·2

The effect of annealing on the properties of the mild steel is clearly illustrated in the above table. There is an increase in ductility accompanied by a marked reduction in the value of the ultimate tensile stress.

45. Proof Stress.—For the higher carbon steels, cast iron and most of the non-ferrous alloys there is no well-defined yielding

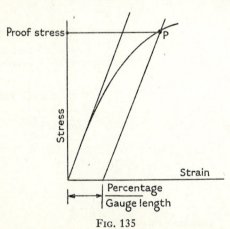

FIG. 135

of the material as in the case of mild steel. The form of the stress-strain diagram for these materials is shown in Fig. 135.

In the absence of a yield stress it is usual to specify a limiting stress called the *proof stress* which corresponds to a definite amount of permanent set. This proof stress figure may be obtained by drawing a line AP from point A on the strain axis so that OA corresponds to the specified percentage of permanent set allowed (suggested figures are 0·1 per cent gauge length for ferrous materials and 0·05 per cent gauge length for non-ferrous materials). AP is drawn parallel to the linear part of the diagram cutting the curved part at P. The stress corresponding to the point P, based on the nominal cross-sectional area of the specimen, gives the required percentage proof stress. The proof stress is usually applied to the specimen for a period of 15 seconds and if, upon its removal, the permanent extension has *not* exceeded the specified percentage of the gauge length, the material is assumed to satisfy the given specification.

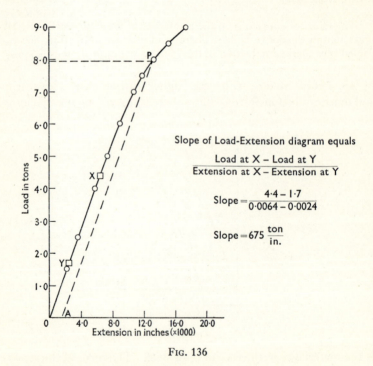

Slope of Load-Extension diagram equals

$$\frac{\text{Load at X} - \text{Load at Y}}{\text{Extension at X} - \text{Extension at Y}}$$

$$\text{Slope} = \frac{4\cdot4 - 1\cdot7}{0\cdot0064 - 0\cdot0024}$$

$$\text{Slope} = 675 \frac{\text{ton}}{\text{in.}}$$

FIG. 136

Example 50

Explain, briefly, the terms elastic limit, permanent set and percentage elongation, as used in connection with tensile testing.

A tensile specimen of a non-ferrous alloy, of diameter 0·564 inch and gauge length 3 inches, gave the following results:

Load, tons	1·5	2·5	4·0	5·0	6·0	7·0	7·5	8·0	8·5	9·0
Extension, inches ×1000	2·2	3·5	5·7	7·3	8·9	10·7	11·8	13·3	15·2	17·4

Fracture occurred at a load of 9·5 tons when the length between the gauge points was 3·18 inches. Draw a graph of these results with the loads as ordinates and determine Young's modulus for the material, the percentage elongation, and the stress at which the permanent set would be 0·05 per cent of the gauge length.   (I. Mech.E.)

Solution

Elastic Limit.—This is the point below which a material remains elastic and will return to its original dimensions on removal of the load.

Permanent Set.—When a material is loaded *beyond* the elastic limit and the load removed it will not return completely to its original dimension. The change in dimensions which exists after the load is removed is called the permanent set.

Percentage Elongation.—This is defined as the increase in the gauge length at fracture expressed as a percentage of the original gauge length.

Young's Modulus.—This may be calculated from the linear part of the diagram which is plotted from the figures given—see Fig. 136.

$$E = \frac{Stress}{Strain} = \frac{Load/Cross\text{-}sectional\ Area}{Extension/Gauge\ length}$$

$$E = \frac{Load}{Extension} \times \frac{Gauge\ Length}{Cross\text{-}sectional\ area}.$$

Hence $E = $ Slope of Load Extension diagram $\times \dfrac{3}{\frac{\pi}{4}(0·564)^2}$.

From Fig. 136 slope of diagram $= 675$ tons/in.

$$\therefore\ E = 675 \times \frac{3}{0·25} = \textbf{8100 tons/in.}^2$$

$$\% \text{ Elongation} = \frac{\text{Increase in gauge length at fracture}}{\text{Original gauge length}} \times 100$$

$$= \frac{(3·18 - 3·00)}{3·00} \times 100 = \textbf{6}\% \text{ (on 3'' G.L.)}$$

0·05 per cent Proof stress. OA on the load-extension diagram is made equal to 0·05 per cent of the 3-inch gauge length, i.e. 0·0015 inch.

AP is drawn parallel to the linear part of the diagram to give the proof load P as 7·95 tons,

$$\therefore \text{ Proof Stress} = \frac{7\cdot 95}{0\cdot 25} = 31\cdot 80 \text{ tons/in.}^2$$

Example 51

A bar, 24 inches long, is made up of a rod of steel, 8 inches long and 1 inch diameter, fixed at one end to a rod of copper 16 inches long. Determine the necessary diameter of the copper rod in order that the extension of each material shall be the same under an axial tensile load of 2 tons. What will then be the stresses in the steel and the copper ?

(E for steel 30×10^6 lb./in.^{2} E for copper 16×10^6 lb./in.^{2})

Solution

Each material will carry the same load, 2 tons. (Fig. 137.)

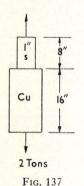

2 Tons

Fig. 137

From $E = \dfrac{\text{Stress}}{\text{Strain}}$, $E = \dfrac{PL}{Al}$ and $l = \dfrac{PL}{AE}$,

where l is the extension and L the original length. Since l for steel $= l$ for copper, we have

$$\frac{PL_s}{A_s E_s} = \frac{PL_c}{A_c E_c}.$$

$$\therefore \frac{8}{\frac{\pi}{4}\cdot 1^2\, 30 \times 10^6} = \frac{16}{\frac{\pi d^2}{4}\cdot 16 \times 10^6},$$

and $d^2 = 3\cdot 75$

$$\therefore \textbf{Dia. of copper} = \textbf{1·94 in.}$$

The stress in the **steel** $= \dfrac{P}{A_s} = \dfrac{2}{\frac{\pi}{4}\cdot 1^2} = \textbf{2·54 tons/in.}^2$

The stress in the **copper** $= \dfrac{P}{A_c} = \dfrac{2}{\frac{\pi}{4}\cdot 1\cdot 94^2} = \textbf{0·68 tons/in.}^2$

Example 52

A compound cylindrical bar is made up of a $1\frac{1}{4}$ inches diameter bar of brass tightly fitting inside a sheath of steel. The steel sheath is to be designed to take $\frac{5}{8}$ of the total axial tensile load. What must be the thickness of the steel sheath, if E for brass $=6500$ tons/in²., and E for steel $=13,500$ tons/in.² ? What will be the stress in each material, if an axial pull of 5 tons is applied to the end of the bar ?

(N.C.T.E.C.)

Solution

It must be assumed that the length of the steel and the brass remain equal. Then the strain in each material must be the *same*. (Fig. 138.)

$\therefore$ Strain in steel $=$ Strain in brass,

or
$$\frac{f_S}{E_S}=\frac{f_B}{E_B}.$$

If W_S and W_B are the loads carried by the steel and the brass and A_S and A_B are their respective cross-sectional areas we have :

$$\frac{W_S}{A_S E_S}=\frac{W_B}{A_B E_B}.$$

But if W is the total load carried by the compound bar, $W_S=\frac{5}{8}W$ and $W_B=\frac{3}{8}W$,

$$\therefore \frac{\frac{5}{8}W}{\frac{\pi}{4}(D^2-1\frac{1}{4}^2)E}=\frac{\frac{3}{8}W}{\frac{\pi}{4}1\frac{1}{4}^2E_B},$$

where D $=$ O.D. of steel sheath.

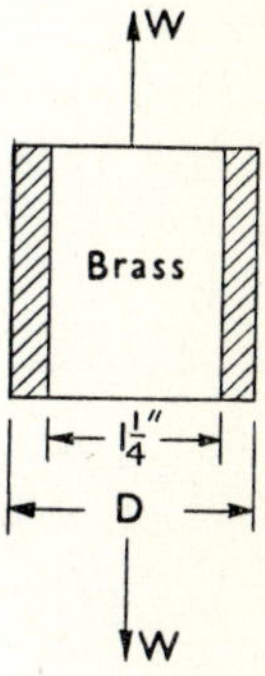

Hence
$$\frac{5}{13500.(D^2-1\frac{1}{4}^2)}=\frac{3}{1\frac{1}{4}^2.6500}.$$

Rearranging
$$(D^2-1\frac{1}{4}^2)405=325.1\frac{1}{4}^2,$$

$$\therefore D=1\frac{1}{4}\sqrt{\frac{730}{405}} \text{ and } D=1\text{·}68 \text{ in.}$$

Therefore, thickness of steel sheath $=\dfrac{1\text{·}68-1\text{·}25}{2}.$

Thickness $=0\text{·}215$ in.

For an axial load   W $=5$ tons.

Fig. 138

Stress in **Steel** $=\dfrac{\frac{5}{8} \cdot 5}{\frac{\pi}{4}(1 \cdot 68^2 - 1 \cdot 25^2)}$ $= 3 \cdot 16$ tons/in.^{2}

Stress in **Brass** $=\dfrac{\frac{3}{8} \cdot 5}{\frac{\pi}{4} \cdot 1 \cdot 25^2}$ $= 1 \cdot 53$ tons/in.^{2}

46. Temperature Stresses.—When a compound bar, Fig. 139, is subjected to a temperature rise the stresses induced in the individual members may be determined as follows.

If we imagine that one of the end fixings is removed and *then* the temperature rise applied the bars will be free to expand as shown at (b).

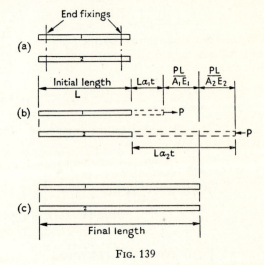

Fig. 139

Clearly, they will expand by differing amounts, depending on their relative coefficients of linear expansion. Now in fact, of course, this free expansion is not allowed because the fixing compels the bars to finish having the same length. It will be evident from (b) that the fixing must supply a force which will extend bar (1) and compress bar (2) until this common length is established at (c).

Once the value of this fixing force is determined the stress in each bar can easily be calculated.

The free expansion of the bars is given by $L\alpha_1 t$ and $L\alpha_2 t$, where t is the given temperature rise.

Due to the fixing force P, the extension of bar (1) will be $\dfrac{PL}{A_1E_1}$ and the compression of bar (2) will be $\dfrac{PL}{A_2E_2}$. (The error introduced by calculating these extensions on the original length L instead of the freely expanded length will be negligible.)

Then from the diagram (b) we have

$$L\alpha_1 t + \frac{PL}{A_1E_1} = L\alpha_2 t - \frac{PL}{A_2E_2},$$

which by rearrangement becomes :

$$L\alpha_2 t - L\alpha_1 t = \frac{PL}{A_1E_1} = \frac{PL}{A_2E_2}.$$

Dividing through by L :

$$\alpha_2 t - \alpha_1 t = \frac{P}{A_1E_1} = \frac{P}{A_2E_2},$$

or   **Difference of Expansions/unit length**
                     = Sum of the fixing strains.

Example 53

A steel bolt, 1 inch diameter, passes through a copper tube, inside diameter $1\frac{1}{2}$ inches and outside diameter 2 inches. The length of the bolt and tube is 36 inches, when the nut is given a quarter of a turn. If the pitch of the bolt thread is $\frac{1}{8}$ inch, determine the stresses produced in the bolt and the tube.

If the temperature of the assembly is now increased by 150° F. what will be the final stresses in each material ?

(Young's Modulus   Steel 30×10^6 lb./in.^{2} ; Copper 16×10^6 lb./in.^{2}

Coefficient of Linear
    Expansion   Steel $6 \times 10^{-6}/°$ F.   Copper $10 \times 10^{-6}/°$ F.)

Solution

Stresses due to tightening the nut

Tightening the nut will put the bolt in tension and the tube in compression.

Let W equal the tensile force in the bolt = Compressive force in the tube.

Then under this force the extension of the bolt $=\dfrac{WL}{A_s E_s}$, and the compression of the tube $=\dfrac{WL}{A_c E_c}$.

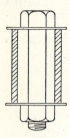

Now the displacement of the nut—which is a quarter of the pitch—will equal the extension of the bolt plus the compression of the tube.

Therefore $\dfrac{WL}{A_s E_s}+\dfrac{WL}{A_c E_c}=\tfrac{1}{4}$ pitch.

$$A_s =\frac{\pi}{4}=0 \cdot 785 \text{ in.}^2\,;\ \ A_c =\frac{\pi}{4}((2)^2 -(1\tfrac{1}{2})^2)=1 \cdot 375 \text{ in.}^2$$

Fig. 140

$$\therefore\ \ \frac{W \cdot 36}{0 \cdot 785 \times 30 \times 10^6}+\frac{W36}{1 \cdot 375 \times 16 \times 10^6}=\tfrac{1}{4}\times\tfrac{1}{8}.$$

$$\therefore\ \ W =9850 \text{ lb.}$$

Then tensile stress in steel $=W/A_s =\dfrac{9850}{0 \cdot 785}=12500 \text{ lb./in.}^2$

Compressive stress in copper $=W/A_c =\dfrac{9850}{1 \cdot 375}=7150 \text{ lb./in.}^2$

Stresses due to increasing the temperature

The stresses set up due to the change in temperature will be in *addition* to those induced by tightening the nut.

Let P equal the force induced in the bolt and the tube due to the temperature increase 't'.

Then since :

Difference of Expansion/unit length $=$ Sum of the strains

$$\alpha_c t - \alpha_s t =\frac{P}{A_s E_s}+\frac{P}{A_c E_c}.$$

$$\therefore\ \ 10^{-6}(10-6)150=\frac{P}{0 \cdot 785 \times 30 \times 10^6}+\frac{P}{1 \cdot 375 \times 16 \times 10^6}$$

$$600 =0 \cdot 0425P +0 \cdot 0455P.$$

$$\therefore\ \ P =6820 \text{ lb.}$$

Tensile stress in steel $=\dfrac{P}{A_s}=\dfrac{6820}{0 \cdot 785}=8700 \text{ lb./in.}^2$

Compressive stress in copper $=\dfrac{P}{A_c}=\dfrac{6820}{1 \cdot 375}=4970 \text{ lb./in.}^2$

Therefore final stress due to tightening nut and temperature change is given by :

Final tensile stress in bolt $=12500+8700=$ **21200 lb./in.²**

Final Compressive stress in tube $=7150+4970=$ **12120 lb./in.²**

47. Cottered Joints.—The problems dealt with so far have involved tensile and compressive stresses only. As an example in which shear stresses must be taken into account, consider the cottered joint shown in Fig. 141.

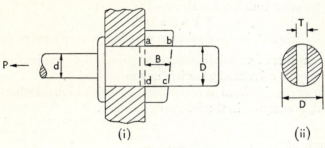

(i)   (ii)

FIG. 141

In the joint illustrated the two members are joined together by means of a tapered cotter which passes through a slot in the rod. The joint supports an axial load P, which produces tensile stresses in the rod, shearing stresses in the cotter and bearing or compressive stresses between the cotter and the rod. The maximum load which can be carried may be determined by considering the possible ways in which elastic failure of the joint can occur.

The joint may fail :

(*a*) In tension across the minimum diameter of the rod—'*d*'.

The area resisting tensile stress $=\dfrac{\pi d^2}{4}$.

Then
$$P = f_t \frac{\pi d^2}{4},$$

where f_t is the maximum permissible tensile stress.

(*b*) In tension across the reduced section of the rod—see Fig. 141 (ii).

The area resisting tensile stress is approximately $\dfrac{\pi D^2}{4} - DT$.

Then
$$P = f_t\left(\frac{\pi D^2}{4} - DT\right).$$

(c) In shear across the *two* cotter sections *ab* and *dc*.

Area resisting shearing stress $= 2 \times B \times T$.
Then
$$P = f_s \times 2BT,$$
where f_s is the maximum permissible shearing stress.

(d) In compression across the bearing surface of the cotter *bc*.

Area resisting compressive stress $= DT$.
Then
$$P = f_c \times DT,$$
where f_c is the maximum permissible compressive stress.

From these four relationships the principal dimensions of the joint may be calculated to satisfy any given load P.

An alternative form of cotter joint, in which both members are slotted, is analysed in the following example.

Example 54

Fig. 142 illustrates a cottered connection between two shafts. The joint is to transmit an axial load of 24 tons, the maximum permissible working stresses being as follows :

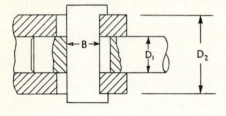

Tension 8 tons/in.²
Shear 6 tons/in.²
Bearing 12 tons/in.²

Calculate approximate values for the dimensions D_1, D_2, B and T. You may ignore the curvature of the shear area, and may assume bearing areas to be rectangular.

(U.E.I.)

Fig. 142

Solution

Tension

The area of the bar resisting failure in tension

$$= \left(\frac{\pi}{4}D_1{}^2 - D_1T\right).$$

Then since $\qquad$ Area $\times f_t =$ Load

$$\therefore \left(\frac{\pi}{4}D_1{}^2 - D_1T\right)8 = 24. \qquad . \qquad . \qquad \text{(i)}$$

Compression

The bearing area of the cotter in the slot of shaft $D_1 = D_1T$ (*ad* in diagram, Fig. 143).

Then $\quad$ Area $\times f_c =$ Load.

$$\therefore \; D_1T \times 12 = 24. \qquad . \qquad \text{(ii)}$$

The bearing area of the cotter in the slot of shaft D_2 is given by $2 \times t \times T$ (*be* and *cf* in diagram).

Hence $\quad 2t \cdot T \times 12 = 24. \qquad . \qquad$ (iii)

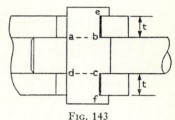

FIG. 143

Shear

The cotter is in double shear cross-sections *ab* and *cd*.

$$\therefore \text{ Area resisting shear} = 2BT.$$

Then $\qquad\qquad\qquad$ Area $\times f_s =$ Load,

or $\qquad\qquad\qquad 2BT \times 6 = 24. \qquad . \qquad . \qquad$ (iv)

From equation (ii) $\qquad DT = 2\text{in.}^2$

Substituting this value in equation (i) we get

$$\left(\frac{\pi}{4}D_1{}^2 - 2\right)8 = 24,$$

$$\therefore \; D_1 = 2\cdot52 \text{ in.}$$

Since $\qquad D_1T = 2 \text{ in.}^2$ and $D_1 = 2\cdot52$ in.

$$T = 0\cdot795 \text{ in.}$$

From equation (iv) $\qquad 2BT \times 6 = 24,$

$$B = \frac{2}{T},$$

$$\therefore \; B = 2\cdot52 \text{ in.}$$

Finally using equation (iii), $\quad 2tT \times 12 = 24,$

$$t = \frac{1}{T},$$

$$\therefore \; t = 1\cdot26 \text{ in.}$$

Then since $D_2 = D_1 + 2t,$

$D_2 = 2.52 + 2 \times 1.26$

and $D_2 = 5.04$ in.

Approximate values of the dimensions are asked for in the question so the required values may be given as :

$D_1 = 2.5$ in. $D_2 = 5.0$ in. $B = 2.5$ in. $T = 0.8$ in.

48. Stresses in Thin Cylindrical Shells.—When a thin cylindrical shell is subjected to internal pressure the cylinder will fail if the stresses set up are allowed to exceed the permissible working stress for the material.

There are two ways in which failure may occur, as illustrated in Fig. 144. The cylinder may burst along a longitudinal seam

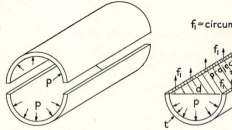
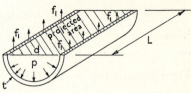

f_1 = circumferential or hoop stress

Failure along longitudinal section
(a)

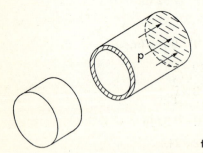
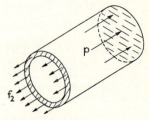

f_2 = longitudinal stress

Failure across transverse section
(b)

FIG. 144

as shown at (a) or it may fail across a transverse section as at (b).

We can examine the stress conditions along any horizontal section by considering the equilibrium of half the cylinder as shown in Fig. 144 (a).

For equilibrium the internal pressure force is balanced by *circumferential* or *hoop* stresses f_1 which act over the area, $2(L \times t)$. (It is usual to ignore the strengthening effects of any support offered by the end plates of the cylinder.)

Therefore, resistance to bursting $= f_1 \times 2(L \times t) = 2f_1 Lt$.

Now the pressure p acts over the horizontal projected area $d \times L$.

Hence for equilibrium $2 f_1 t L = p.d.L$

or $$f_1 = \frac{pd}{2t}.$$.   .   .   (1)

Then thickness of plates $$t = \frac{pd}{2f_1}.$$.   .   (2)

Equation (2) gives the required thickness of plates for a given material strength f_1. Where the cylinder is made up of plates riveted together—as in the case of a boiler shell—Fig. 145, the

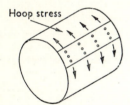

Hoop stress

Rivets on longitudinal joint

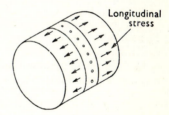

Longitudinal stress

Rivets on circumferential joint

FIG. 145

strength of the joint is always less than that of the solid plate. To allow for this the thickness of plate is increased beyond that given in equation (2) by employing the relationship :

Strength of joint = Efficiency of Joint (η) × Strength of Solid Plate.

Then if f_1 is taken as the permissible stress for the solid plate we have

$$t = \frac{pd}{2f_1 \times \text{Joint Eff.\%}}, \text{ or } t = \frac{pd}{2f_1\eta}.$$

A.M.—N

For equilibrium on a transverse section, Fig. 144 (b), the pressure p acting on the area of the end plates $\frac{\pi d^2}{4}$ is balanced by the longitudinal stresses f_2 on a transverse area approximately equal to πdt.

Hence
$$f_2 \pi dt = \frac{p\pi d^2}{4},$$

and
$$\mathbf{f_2} = \frac{\mathbf{pd}}{\mathbf{4t}}. \qquad . \qquad . \qquad . \qquad (3)$$

A comparison of equations (1) and (3) will indicate that the *hoop* stress on a *longitudinal* section is double the *longitudinal* stress on a *circumferential* section. From this it is evident that the longitudinal joints must be made much stronger than the circumferential joints (Fig. 145)—the thickness of plates being calculated from equation (2).

Example 55

Deduce expressions, stating the assumptions made, for the stresses produced in the longitudinal section (hoop stress) and the transverse section (axial stress) of a thin cylindrical shell subjected to an internal pressure.

A cylindrical shell 8 feet diameter is constructed of mild steel plate $1\frac{1}{8}$ inches thick. The ultimate tensile stress of the steel is 30 tons/in.^{2} Assuming a factor of safety of 6 and that the efficiency of the longitudinal joint is 78 per cent, calculate the maximum safe internal pressure in pounds per square inch gauge to which the shell can be subjected.
(U.L.C.I.)

Solution

The expressions required are those deduced in the previous article, viz. Hoop stress $f_1 = \frac{pd}{2t}$; Longitudinal (Axial) stress $f_2 = \frac{pd}{4t}$.

The assumptions made in deriving these expressions :

(i) the thickness of the shell is small compared with the over-all dimensions allowing the stress on any section to be considered as uniformly distributed.

(ii) the end plates give no support to the sides of the cylinder.

The working stress of the material

$$f_1 = \frac{\text{U.T.S.}}{\text{F. of S.}} = \frac{30}{6} = 5 \text{ tons/in.}^2$$

Then since $d = 8$ ft. $= 96$ in., $t = 1\frac{1}{8}$ in. and $\eta = 0.78$, we have $f_1 = \dfrac{pd}{2t\eta}$, where p is the gauge pressure in ton/in.^{2}

$$\text{i.e. } \quad 5 = \frac{p \times 96}{2 \times 1\frac{1}{8} \times 0.78}.$$

Hence $\qquad\qquad p = 0.091$ ton/in.^{2}

or $\qquad\qquad$ **p = 204 lb/in.^{2}** (max. safe internal pressure).

49. Riveted Joints.—The following notes are meant to serve as a reminder of the principles involved in determining the strength of riveted joints only in so far as they are required in 'third year' calculations. For a full discussion of the proportion of riveted joints the student is referred to books on Machine Design.*

Fig. 146 illustrates the two main types of joint in use. In the lap joint shown at (*a*) the two plates lap over each other and are connected by one or more rows of rivets. To break this joint by

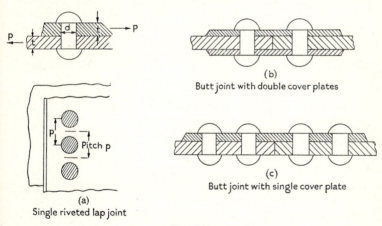

(b)
Butt joint with double cover plates

(c)
Butt joint with single cover plate

(a)
Single riveted lap joint

Fig. 146

shearing the rivets, each rivet has to shear across one section only. In the butt joint with double cover plates (*b*) it should be observed that each rivet is in *double* shear, i.e. the area resisting shear

* *Machine Design*, Unwin (Longmans) ; *Intermediate Eng. Dwg.*, Parkinson (Pitmans).

equals $2 \cdot \frac{\pi d^2}{4}$. As a precaution against uneven bearing of the rivet in the double cover plates the shear area for this joint is often taken as only $1 \cdot 75 \frac{\pi d^2}{4}$.

The single cover plate butt joint at (c) consists of a lap joint between the cover plate and each member.

In the design of any joint we require to investigate the resistances of the joint to failure in tension, shear and compression.

Consider one pitch width of the single riveted lap joint in Fig. 147.

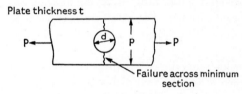

Plate thickness t

Failure across minimum section

FIG. 147

If the plate fails in tension it will clearly give way across the section through the rivet hole.

The area of metal resisting tension failure $= (p - d)t$, and if f_t is the permissible tensile stress in the plate the maximum load is given by

$$P = f_t(p - d)t. \qquad . \qquad . \qquad . \qquad (1)$$

If failure occurs due to shearing of the rivet across an area of rivet equal to $\frac{\pi d^2}{4}$ and f_s is the permissible shearing strength of the rivet, we have

$$P = f_s \frac{\pi d^2}{4}. \qquad . \qquad . \qquad . \qquad (2)$$

Finally, failure may be due to crushing of the rivet between one of the plates. The projected area of rivet resisting crushing is $d \times t$, so if f_c is the permissible bearing stress on the rivet,

$$P = f_c \cdot dt. \qquad . \qquad . \qquad . \qquad (3)$$

Now if the rivet diameter is determined according to Unwin's empirical equation $d = 1 \cdot 2\sqrt{t}$, where t is the plate thickness, the joint will not fail in compression. On this basis the joint is then made equally strong against tension and shear failure by equating the values of P given in equations (1) and (2).

Then

$$f_t(p - d)t = f_s \frac{\pi d^2}{4}.$$

This equation will give the pitch of the rivets in terms of the rivet diameter.

Thus

$$p = \frac{f_s}{f_t} \frac{\pi d^2}{4t} + d.$$

Joint Efficiency.—When a rivet hole is cut through the plate in a single riveted lap joint the strength of each pitch width of plate is reduced from its original value $f_t p t$ to the value $(p - d)t f_t$. The ratio of this latter value to the strength of the original plate is called the tearing efficiency of the joint.

Hence tearing efficiency $= \dfrac{(p - d)tf_t}{ptf_t} = \dfrac{p - d}{p}.$

Similarly,

Shearing efficiency $= \dfrac{\text{Joint resistance to shear}}{\text{Strength of original plate}} = \dfrac{\frac{\pi d^2}{4}f_s}{ptf_t},$

and   Crushing efficiency $= \dfrac{\text{Joint resistance to crushing}}{\text{Strength of original plate}} = \dfrac{dtf_c}{ptf_t}.$

The efficiency values of lap joints with more than one row of rivets and also for butt joints may be obtained in a similar manner.

Example 56

A thin cylindrical boiler shell is 4 feet diameter and $\frac{1}{2}$ inch thick. The longitudinal joint is a single-riveted lap joint having rivets $\frac{7}{8}$ inch diameter spaced at intervals of $3\frac{1}{4}$ inches. The maximum allowable stresses are 6 ton/in.^{2} in tension, 5 tons/in.^{2} in shear and 12 tons/in^2 in bearing. Determine :

(a) the efficiency of the joint, and
(b) the greatest internal pressure the boiler shell can safely carry.
(U.E.I.)

Solution

(a) Efficiency of joint (η). All three forms of failure must be considered as follows :

Tearing efficiency $= \dfrac{p - d}{p} = \dfrac{3\frac{1}{4} - \frac{7}{8}}{3\frac{1}{4}}$

Tearing efficiency $= 73\%.$

Shearing efficiency $= \dfrac{\pi d^2 f_s}{4pt f_t} = \dfrac{\pi(\frac{7}{8})^2 5}{4 \times 3\frac{1}{4} \times \frac{1}{2} \times 6}$

Shearing efficiency $= 31\%$.

Crushing efficiency $= \dfrac{df_c}{pf_t} = \dfrac{\frac{7}{8}}{3\frac{1}{4}}\dfrac{12}{6}$

Crushing efficiency $= 54\%$.

Hence, the efficiency of the joint will be the least of these values, or

Joint efficiency $\eta = 31\%$.

(b) Maximum internal pressure (p).

From the equation $f_t = \dfrac{pd}{2t\eta}$, where d is now the internal dia. of the shell.

we have $\qquad p = \dfrac{2t\eta}{d} f_t = \dfrac{2 \times \frac{1}{2} \times 0 \cdot 31 \times 6}{48}.$

$\therefore$ **p $= 0 \cdot 039$ ton/in.² or p $= 87$ lb./in.²**

Example 57

Two lengths of mild steel tie bar, $9\frac{1}{2}$ inches wide and $\frac{5}{8}$ inch thick, are to be joined by a double strapped butt joint using mild steel rivets. Determine the number of rivets required and indicate a suitable arrangement for them. The following assumptions may be made :

Rivet dia. $d = 1 \cdot 2\sqrt{t}$; Rivet pitch $= 3d$; Factor for double shear $= 1 \cdot 75$.

Also calculate the maximum load which can be carried by the joint and the joint efficiency. The maximum permissible stresses in tension, shear and bearing are 7 tons/in.², 5 tons/in.² and 12 tons/in.² respectively.

Solution

From the data given, rivet dia. $d = 1 \cdot 2\sqrt{t}$, where t is the plate thickness.

$d = 1 \cdot 2\sqrt{0 \cdot 625} = 0 \cdot 95$ in.   Use rivets 1 in. dia.

then rivet pitch $= 3d = 3$ in.

The shear strength of 1 rivet

$= \dfrac{\pi d^2}{4} 1 \cdot 75 f_s = \dfrac{\pi}{4} \times 1 \times 1 \cdot 75 \times 5 = 6 \cdot 875$ tons.

The bearing strength of 1 rivet

$= d \cdot t f_c = 1 \times \frac{5}{8} \times 12 = 7 \cdot 5$ tons.

Since the bearing resistance is greater than the shearing resistance, the number of rivets will be determined on the basis of equal strength in tension and in shear.

Assume that the full load will be carried by the width of plate less one rivet (diameter) hole.

Then the net area of plate supporting tension $=(9\frac{1}{2}-1)\frac{5}{8}$, and tensile load $=8\frac{1}{2}\times\frac{5}{8}\times 7 =$ **37·2 tons.**

Now on the basis that strength in tension = strength in shear, we get tensile load = Shear strength of 1 rivet × No. of rivets,

or $\qquad 37\cdot2 = 6\cdot875 \times N$

$\qquad N = 5\cdot41$, say 6 rivets on each side of joint.

A suitable arrangement of 6 rivets is shown in Fig. 148.

To determine the maximum load and efficiency we must now investigate the load carrying capacity of the arrangement suggested. Consider each row of rivets in turn :

Across section AA : Safe load as calculated = **37·2 tons.**

Across section BB : Failure across BB is resisted by the strength of the reduced width of plate at this section plus the shearing resistance of the single rivet to the left of BB.

Hence safe load at BB

$$= (9\cdot5 - 2\times 1)\times\frac{5}{8}\times 7 + 6\cdot875$$
$$= 32\cdot85 + 6\cdot875 = \textbf{39·725 tons.}$$

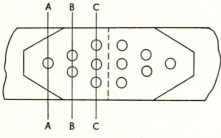

Fig. 148

Across section CC : Safe load = strength of reduced width at CC plus shearing resistance of 3 rivets to left of CC.

Safe load at CC $= (9\cdot5 - 3\times 1)\times\frac{5}{8}\times 7 + 6\cdot875\times 3$
$\qquad\qquad = 28\cdot4 + 20\cdot6 = \textbf{49·0 tons.}$

The minimum strength is that calculated for section AA.

Hence $\qquad\qquad$ Maximum load = **37·2 tons.**

The efficiency of joint $= \dfrac{\text{Strength of weakest section}}{\text{Strength of original plate}}$.

Now strength of original plate

$$= \text{Area plate} \times f_t = 9\tfrac{1}{2} \times \tfrac{5}{8} \times 7 = \textbf{41·56 tons.}$$

$$\therefore \ \text{Eff. } \% = \frac{37\cdot2}{41\cdot56} \times 100 = \textbf{89·4}\%.$$

50. Strain Energy.—When a bar is elongated by a gradually increasing tensile load work is done on the bar which is stored up in the bar in the form of *strain energy*. If the strain is kept within the elastic range this energy can be recovered by gradually unloading the bar.

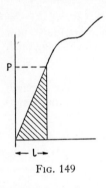

FIG. 149

Consider the load-extension diagram in Fig. 149. '*l*' is the extension associated with a gradually applied load of final magnitude P. The shaded area of the triangle in this diagram gives the work done in increasing the load from zero to the final value P. This work done represents the total strain energy U, which is stored up in the bar during straining and is given by

$$U = \tfrac{1}{2}Pl.$$

Therefore,   **Work Done = Strain Energy $= \tfrac{1}{2}Pl$.**

If f is the tensile stress corresponding to the load P, A the cross-sectional area of the bar and L its original length, then $\dfrac{P}{A} = f$ and $P = Af$.

Also since Strain = Stress/E, $l = \dfrac{fL}{E}$.

Substituting for P and l in the equation for U, we get

$$U = \frac{f^2}{2E} . AL.$$

But AL is the original volume of the bar,

$$\therefore \ U = \frac{f^2}{2E} \times \text{Volume of bar}$$

or
$$U = \frac{f^2}{2E} \text{ per unit volume.} \qquad . \qquad . \qquad (1)$$

Strain energy per unit volume is often called *resilience*. The greatest amount of strain energy which can be stored in a bar without permanent set will be given when f in equation (1) is equal to the stress at the elastic limit (or the proof stress, whichever is applicable).

This particular value of the strain energy is called the *proof resilience*.

Example 58

Define Resilience and Proof Resilience.

Derive from first principles an expression giving the resilience of a bar in tension.

A piece of steel is 4 inches × 6 inches in cross-section and 20 inches long. Find the resilience of the bar when carrying a load of 48 tons. What is the maximum amount of energy which can be stored in this piece, if the elastic limit load is 360 tons and the modulus of elasticity is 13,500 tons/in.^{2}? (N.C.T.E.C.)

Solution

$$\text{Resilience} = \text{Strain Energy} = \frac{f^2}{2E} \times \text{Volume.}$$

Since
$$f = \frac{P}{A} \text{ and Volume} = AL$$

$$\text{Resilience} = \frac{P^2}{2EA^2}AL = \frac{P^2L}{2AE}.$$

For $P = 48$ tons
$$\text{Resilience} = \frac{(48)^2 20}{2 \times 13500 \times 4 \times 6} = 0.071 \text{ in. ton.}$$

For $P = 360$ tons

$$\begin{matrix} \text{Proof Resilience or} \\ \text{Max. Strain Energy} \end{matrix} = \frac{(360)^2 20}{2 \times 13500 \times 4 \times 6} = 4.0 \text{ in. tons.}$$

51. Stress due to Impact.—A given load will induce a greater stress in any structure when applied impulsively than when applied gradually.

This may be demonstrated by means of the arrangement shown in Fig. 150. A weight W is allowed to fall through a height h onto a stop at the end of the bar fixed at its upper end and the resulting impact produces an elongation of the bar.

As the bar extends the velocity of the weight at impact will be reduced by the opposing resistance of the bar. At the instant this velocity is zero the extension will cease and the tensile stress induced in the bar will have reached its maximum value.

The initial effect on the bar of the sudden impact will be to cause it to oscillate like a spring, but it will finally settle down with an extension equal to that for the load W, if gradually applied.

On the assumption that there is no loss of energy at impact the total work done by the weight will be transformed into strain energy in the bar.

If x is the maximum elongation, the weight will fall through a total height of $(h + x)$.

Then work done by $W = W(h + x)$.

If f is the maximum stress induced in the bar,

then
$$U = \frac{f^2}{2E} \times AL.$$

This strain energy may be expressed in terms of the extension

Area A

W

L

h

FIG. 150

x by means of the relationship $f = \frac{x}{L} \cdot E$.

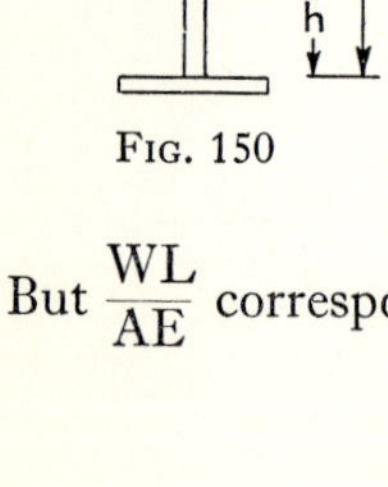

Then
$$U = \left(\frac{xE}{L}\right)^2 \frac{AL}{2E} = \frac{x^2 EA}{2L}.$$

Now, equating work done to strain energy,

$$W(h + x) = \frac{x^2 EA}{2L}.$$

Transposing
$$\frac{WL}{AE}(h + x) = \frac{x^2}{2}$$

But $\dfrac{WL}{AE}$ corresponds to the extension for W gradually applied,

i.e. $\dfrac{WL}{AE} = l$ the static extension.

$$\therefore \ l(h + x) = \frac{x^2}{2}.$$

This gives a quadratic equation in x—the maximum extension due to an impulsive load W.

Hence   maximum $x = l + \sqrt{l^2 + 2hl}$.   .   .   (1)

With x thus determined the maximum stress is easily obtained from

$$f = \frac{x}{L} \cdot E. \qquad \qquad \qquad (2)$$

Clearly if the height h is *large* compared with the static extension l equation (1) reduces to :

$$x = \sqrt{2hl}.$$

Then from (2) $\qquad \qquad f = \sqrt{2hl} \cdot \frac{E}{L},$

or $\qquad \qquad \qquad f = \sqrt{\frac{2hWE}{AL}}.$

These formulae need not be remembered.

52. Suddenly Applied Load.—If h is zero, i.e. if W is suddenly applied to the stop, there will be no impact velocity. To determine the stresses induced in the bar for this case consider the equation

$$x = l + \sqrt{l^2 + 2hl}.$$

If h is zero $\qquad \qquad x = 2l.$

Clearly, this indicates that the initial extension due to W suddenly applied is *twice* as great as that obtained when W is gradually applied. It follows that the stress induced will also be *twice* as great.

Example 59

Explain the meaning of resilience in connection with the straining of a bar.

A steel specimen was turned to a diameter of 0·564 inch and the distance between the gauge points was 8 inches. On being tested in tension the elongation for a load of 2·5 tons was 0·006 inch. Determine the modulus of elasticity of this steel.

A weight of 30 lb. falls through a height of 5 inches, and then commences to stretch a bar of this steel 0·75 inch diameter and 4 feet long. Assuming the elongation produced is small compared with the height through which the weight falls, calculate (*a*) the maximum stress induced in the bar, (*b*) the maximum elongation of the bar.

(U.L.C.I.)

Solution.   See article 50.

To determine E,

$$\text{Stress for 2·5-ton load} = \frac{2·5}{\frac{\pi}{4}(0·564)^2} = 10 \text{ tons/in.}^2$$

$$\text{Strain on 8 in. gauge length} = \frac{0·006}{8} = 0·00075$$

$$E = \frac{\text{Stress}}{\text{Strain}} = \frac{10}{0·00075} = \mathbf{13350 \text{ tons/in.}^2}$$

Maximum Stress.   On the assumption stated,

$$\text{Max. Stress } f = \sqrt{2h \cdot \frac{WE}{AL}}$$

$h = 5$ in.   W = 30 lb.   L = 48 in. $A = \frac{\pi}{4} \cdot (\tfrac{3}{4})^2$ in.^{2} = 0·442 in.^{2}

$$\therefore f = \sqrt{\frac{2 \times 5 \times 30 \times (13350 \times 2240)}{0·442 \times 48}},$$

f = 20600 lb./in.^{2}

Maximum Elongation, from Stress = Strain × E.

$$f = \frac{x}{L} \times E, \text{ where } x \text{ is the maximum elongation.}$$

Then
$$x = \frac{fL}{E} = \frac{20600 \times 48}{13350 \times 2240}.$$

Maximum Elongation = 0·033 in.

Example 60

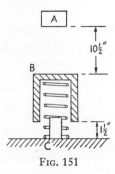

A B

$10\tfrac{1}{2}''$

$1\tfrac{1}{2}''$

C

Fig. 151

A weight of 20 lb. falls through a height of 10·5 inches before striking the spring-loaded cylinder C shown in Fig. 151. The cylinder is 9 inches long over-all and 3 inches external diameter. It has a central hole 2 inches diameter bore by 7 inches deep. The close coiled helical compression spring has a stiffness of 200 lb. per inch and the weight causes the spring to compress and the lower edge of the cylinder to just strike the plate C. Neglecting the compression of the cylinder, deformation of the plate and loss of energy due to impact, determine the maximum compressive stress induced in the cylinder due to impact. $E = 30 \times 10^6$ lb./in.²

Solution

For the conditions stated, the work done by the falling weight will equal the work done in compressing the spring plus the strain energy stored in the cylinder.

Cylinder

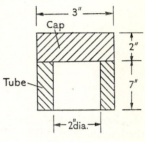

Consider the cylinder in two parts, as shown in Fig. 152, and let

f_1 = Compressive stress in cap
f_2 = Compressive stress in tube.

Then since the load taken by each part is the same :

$$f_1 \frac{\pi}{4}(3)^2 = f_2 \frac{\pi}{4}(3^2 - 2^2) \qquad \therefore \; 9f_1 = 5f_2.$$

FIG. 152

Hence f_2 is the maximum stress in the cylinder and $f_1 = \frac{5}{9}f_2$.

The total strain energy in the cylinder = strain energy in cap + strain energy in tube.

i.e. $U = \dfrac{f_1^2}{2E} \times$ Volume of cap $+ \dfrac{f_2^2}{2E} \times$ Volume of tube.

$$= \frac{(\frac{5}{9}f_2)^2}{2E} \cdot \frac{\pi}{4}3^2 \times 2 + \frac{f_2^2}{2E} \frac{\pi}{4}(3^2 - 2^2) \times 7,$$

or $\qquad\qquad U = \dfrac{16}{E}f_2^2$ in.lb.

Spring

The work done in compressing the spring $= \frac{1}{2}kx^2$, where k = spring stiffness, x = spring compression.

NB.—Initial load on spring = 0.   Final load on spring = kx.

WD = Av. Load $\times$ Distance, i.e. WD $= \dfrac{kx}{2} \cdot x = \dfrac{kx^2}{2}$.

Therefore $\qquad\qquad$ WD $= \frac{1}{2} \cdot 200(1 \cdot 5)^2$.

Work Done = 225 in.lb.

Weight

Work done by weight $= 20 \times (10\frac{1}{2} + 1\frac{1}{2}) = 240$ in.lb.

Then from above

$$\text{WD by weight} = \text{WD on Spring} + U,$$

or
$$240 = 225 + \frac{16f_2{}^2}{30 \times 10^6}$$

$$f_2 = \sqrt{\frac{15 \times 30 \times 10^6}{16}}$$

$$\mathbf{f_2 = 5300 \ lb./in.^2}$$

EXERCISES VI

1. State 'Hooke's Law' and describe an experiment to support the truth of the Law. What is understood by the term 'Young's Modulus of Elasticity'? Explain how this quantity can be determined.

A link in the mechanism of a machine has a cross-sectional area of 0·6 in.^{2} and is 6 feet long. If it carries a tensile load of 3 tons calculate:

(*a*) its elongation under load;

(*b*) its factor of safety, if the ultimate tensile strength is 28 ton/in.^{2}

Take E as 29×10^6 lb./in.^{2} I.Prod.E.

2. Describe, with the aid of a diagram, the principle of operation of any extensometer with which you are familiar and state clearly the precautions that must be observed in its use. In a test on a steel specimen, 0·564 inch diameter, the extensions measured on a gauge length of 2 inches under varying loads are given in the table below. Plot the graph of load against extension and from it deduce the value of the modulus of elasticity of the steel.   U.E.I.

Load ton	0·5	1·0	1·5	2·0	2·5	3·0	3·5
Extension 0·001 inch	0·4	0·7	1·02	1·28	1·60	1·80	2·3

3. A conductor of a transmission system consists of a steel wire, 0·14 inch diameter, covered with an envelope of copper so that the outer diameter is 0·2 inch. The copper adheres to the steel under all conditions of loading. Calculate the stress in the steel and in the copper when a pull of 500 lb. is applied to the conductor and also determine the extension taking place. Also calculate the strain energy of the steel core.

(E steel $= 30 \times 10^6$ lb./in.^{2} E copper $= 18 \times 10^6$ lb./in.^{2})

4. A brass rod, 1 inch diameter, is fitted into a steel tube, 1 inch internal diameter and $1\frac{1}{2}$ inches external diameter. The tube and rod are of the same length. Calculate the maximum compressive load which can be applied to the compound bar if the compressive stresses in the steel and brass must not exceed 6 tons/in.² and 3 tons/in.² respectively.

For brass $E = 12 \times 10^6$ lb./in.²   For steel $E = 30 \times 10^6$ lb./in.²

5. A steel bar, $\frac{3}{4}$ inch diameter and 12 inches long, is placed inside a copper tube, external diameter 1 inch and $\frac{3}{4}$ inch internal diameter. The combination is subjected to an axial thrust of 5 tons. The modulus of elasticity of the metal of the tube is $11\cdot5 \times 10^6$ lb./in.² and of the steel 30×10^6 lb./in.² Find (a) the stress in the tube and the rod, (b) the shortening of the rod, (c) the work done in compression.

6. Define the terms : (a) 'limit of proportionality', and (b) 'modulus of elasticity'.

A pile used for the foundation of a building is made from a cast-iron cylinder, 12 inches outside diameter, thickness 1 inch. The inside of the cylinder is filled with concrete. If the load carried by the pile is 100 tons, find the stresses in the cast iron and the concrete.

(Young's modulus of elasticity for cast iron $= 12 \times 10^6$ lb./in.²
Young's modulus of elasticity for concrete $= 3 \times 10^6$ lb./in.²)

N.C.T.E.C.

7. A $\frac{3}{4}$ inch steel bolt is passed through a brass sleeve of $\frac{7}{8}$ inch bore and $1\frac{1}{8}$ inches outside diameter. A nut and washer are put on the bolt and the nut tightened until the brass sleeve has shortened 0·002 inch. Neglecting any compression of the washer, determine the extension of the bolt.

(E for steel 30×10^6 lb./in.²   E for brass 12×10^6 lb./in.²)

8. A rod in a mechanism is subjected to a maximum tensile force of 5 tons. It is formed of a steel rod, 30 inches long and $1\frac{1}{4}$ inches diameter, fitted with a brass sleeve over the central 20 inches of length. The sleeve is $1\frac{1}{4}$ inches internal diameter, $2\frac{1}{4}$ inches external diameter and is rigidly connected to the steel rod so that both strain together under direct force. The tensile force is applied at the ends of the steel rod. Compare the stress in the steel within the sleeved length with that at the ends, and evaluate the stress in the brass sleeve. Calculate also the total extension produced by the given force. Young's modulus for steel 30×10^6 lb./in.² ; for brass 12×10^6 lb./in.²   I.Mech.E.

9. Working from first principles, find an expression for the stress p set up in a bar of metal whose ends are rigidly held whilst the temperature changes t degrees.

A steel rod, $\frac{1}{2}$ inch diameter, forming part of the construction of a furnace, may be regarded as being held rigidly at its ends. If, under working conditions, the temperature changes 200° F., determine :

(a) the strain set up in the rod ;

(b) the corresponding stress.

The coefficient of linear expansion is 7×10^{-6} per degree F.

Take E as 30×10^6 lb./in.²   I.Prod.E.

10.   A temperature control element using a ¼ inch diameter brass rod 6 inches long is shown in Fig. 153.

Faces A and B are rigidly fixed. There is a gap of 0·003 inch between one end of the rod and face B, when the temperature is 70° F. The other end is fixed rigidly to face A. Calculate :

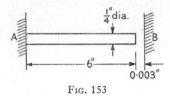

(a) the temperature increase to just close the gap ;

(b) the stress set up in the rod if the temperature increases to 600° F.

Fig. 153

For brass, take coefficient of linear expansion $= 10 \times 10^{-6}/°$ F.

$(E = 13\cdot4 \times 10^6$ lb./in.²)   I.Prod.E.

11.   Two tubes, one of copper and one of steel, are of equal length and rigidly connected together at their ends, so that under all conditions they are of equal length.

The copper tube has internal and external diameters of 4 and 5 inches respectively, whilst the internal and external diameters of the steel tube are 3 and 4 inches respectively. If the original length of the tubes was 15 inches, calculate the stresses set up in them when there is a temperature rise of 40° F.

What is the final length of the tubes ?

(E steel 30×10^6 lb./in.²   E copper 16×10^6 lb./in.²)

Coefficient of linear expansion per ° F., Steel 0·000006 ; Copper 0·0000095.   U.E.I.

12.   A compound bar is made up of a gunmetal rod, 1½ inches diameter, which passes through a steel tube, 2½ inches outside diameter. The rod and tube are firmly fixed together and the compound bar is subjected to a temperature rise of 170° F. Determine the stresses in each material.

It is then required that the stress be removed from the inner rod and a load is, therefore, applied to the steel tube. What is the final stress in the steel tube for this condition ?

Take α for steel $= 6 \times 10^{-6}/°$ F.   E steel $= 30 \times 10^6$ lb./in.²

α for gunmetal $= 11 \times 10^{-6}/°$ F.   E gunmetal $= 13 \times 10^6$ lb./in.²

13.   A brass rod 1 inch diameter is enclosed in a steel tube, 2 inches outside diameter and 1 inch inside diameter, and they are both 4 feet

long at a temperature of 15° C. If the materials are fastened together by two rivets, one at either end, find the fixing force supplied by each rivet if the temperature is raised to 135° C. If the safe double shear stress in each rivet is 8 tons/in.², calculate the necessary diameter of each rivet.

(E for brass 14×10^6 lb./in.²
    Coefficient of expansion, brass $18\cdot6 \times 10^{-6}/°$ C.
    E for steel 30×10^6 lb./in.²
    Coefficient of expansion, steel $11\cdot6 \times 10^{-6}/°$ C.)

14. Derive expressions for the longitudinal and circumferential stresses in a thin boiler shell of thickness t subjected to an internal steam pressure p. A cylindrical boiler, 6 feet diameter, is constructed of plate, $\frac{1}{2}$ inch thick. Assuming the efficiency of the longitudinal riveted joint to be 70 per cent and the safe working stress in the plate to be 5 tons/in.², calculate the greatest allowable pressure in the boiler.

U.E.I.

15. Obtain expressions for the hoop and axial stresses induced in the material of a thin cylindrical shell subjected to an internal pressure.

A boiler shell, 6 feet mean diameter, is constructed of steel plate having an Ultimate Tensile Strength of 28 tons/in.² It is subjected to an internal pressure of 250 lb./in.² gauge. Calculate the thickness of the shell plates, assuming a factor of safety of 5. The holes for the rivets of the longitudinal joints reduce the area of steel under stress by 20 per cent.

16. A thin cylinder, of diameter d and thickness of wall t, is subject to an internal pressure p. Establish formulae giving the circumferential and longitudinal stresses in the wall.

A steel pipe, 20 inches diameter and $\frac{1}{2}$ inch thick, is closed at the ends by bolted flanges and used as a storage vessel for a fluid at a pressure of 200 lb./in.² Connections to the pipe necessitate, in places, the use of longitudinal joints, the efficiency of which may be taken at 80 per cent. Each end flange is secured with 14 bolts which have a diameter, at the bottom of the thread, of 1·15 inches. To ensure joint tightness the bolts are screwed up to give a stress 40 per cent greater than that due to the pressure. If the pipe and bolt material has an ultimate strength of 30 tons/in.², compare the factors of safety for the wall and the bolts. I.Mech.E.

17. A tie member in a roof truss has to carry an axial load of 48 tons. The member is a flat bar $\frac{5}{8}$ inch thick and of constant width. Design and sketch a double covered butt joint in this member given that $\frac{7}{8}$ inch diameter rivets are to be used throughout.

Maximum tensile stress, 10 tons/in.²
Maximum bearing stress, 12 tons/in.²
Maximum shear stress, 6 tons/in.²   U.E.I.

A.M.—O

18. Explain the term resilience and show that for a uniform bar of elastic modulus E and carrying a longitudinal stress f the resilience per unit volume is given by $U = f^2/2E$.

A steel bar, 3 feet long, is 2 inches diameter for 16 inches of its length and 3 inches diameter for the remainder. Calculate the maximum resilience of the bar, if the maximum allowable stress is limited to 10 tons/in.²

Assume $E = 13,500$ tons/in² U.E.I.

19. A reinforced concrete column is 10 feet high and of uniform cross-section 15 inches × 15 inches. It is reinforced by four 1 inch diameter steel rods, symmetrically placed.

If the column carries an axial load of 60 tons, determine the stresses in the steel reinforcement and in the concrete. How much energy is stored in the column ?

(E (steel) $= 30 \times 10^6$ lb./in.² E (concrete) $= 2 \times 10^6$ lb./in.²)

U.E.I.

20. A cantilever frame is pinned to a rigid vertical column at A and at B vertically below A. The top member, AC, is horizontal ; it is 75 inches long and has a cross-sectional area of 1 in.² There are two equal members, AD and CD, each 45 inches long and 2 in.² in cross-section. The remaining member, DB is 75 inches in length and 2·5 in.² in cross-sectional area. A downward vertical load of 5 tons acts at C. The frame is pin-jointed throughout. Determine the forces in the members and the total strain energy content of the frame.

(Young's Modulus, 12,000 tons/in.²) I.Mech.E.

21. A cast iron bar of square section, 2 inches side, is strengthened by the addition of a steel plate, 2 inches wide by $\frac{3}{8}$ inch thick, to each of two opposite sides. The plates are connected rigidly to the bar by fitted pins near the ends. The combination, which has a length of 70 inches, is subject to a total tensile force of 18 tons uniformly distributed over the whole area of the end sections. Taking Young's modulus for steel at 13,000 tons/in² and for cast iron at 7000 tons/in.² determine :

(a) the stresses in the two materials and the extension of the bar ;
(b) the strain energy content of each material ;
(c) the forces to be transmitted through the fixing pins.

I.Mech.E.

22. (a) Deduce an expression for the resilience per unit volume of a uniform cylindrical bar having an elastic modulus E and carrying a tensile stress f.

(b) Use this expression to show that a load applied suddenly to a tension member will induce in it a stress twice as great as that which would be set up if the same load were applied gradually. U.E.I.

23. Define the terms 'resilience' and 'proof resilience'. In the case of a uniform bar subjected to a tensile load determine an expression for one of these quantities.

A link in a machine mechanism is 0·5 inch diameter and 13·5 inches long. If the tensile stress at the elastic limit is 40,000 lb./in.^{2} :

(*a*) calculate the proof resilience ;

(*b*) find the force suddenly applied to the link which would produce this resilience ;

(*c*) write down the value of the gradually applied force which would produce the same resilience, then determine the elongation this force would produce.

Take E as 30×10^6 lb./in.^{2} I.Prod.E.

24. During a test on a $\frac{1}{2}$ inch diameter mild steel rod, it was found that the gradual application of a 1 ton load produced an extension of $\frac{1}{8}$ inch. What would be the initial extension of an exactly similar bar hanging vertically, if a load of 150 lb. was dropped from 3 inches on to a collar securely fixed to the lower end ? Find also the maximum stress set up in the rod.

25. Derive an expression for resilience in a body, in terms of its dimensions, the stress induced and the modulus of elasticity of the material.

A vertical rod, 1 inch diameter and 5 feet long, is securely fixed at its upper end. The lower end carries a stop. What is the maximum free height through which a weight of 50 lb. may be allowed to fall onto the stop if the instantaneous stress is not to exceed 6 tons/in.^{2} ?

(E = 13,000 tons/in.^{2})

Chapter VII

BENDING MOMENT AND SHEARING FORCE

53. Bending Moment and Shearing Force Defined.—A beam resting on two supports, as shown in Fig. 154, is easily recognised from our previous work on statics as the particular case of a body at rest under the action of parallel forces.

FIG. 154	FIG. 155

If the beam has a constant cross-sectional area we may consider its weight to be distributed uniformly throughout its length and represent the loading as in Fig. 155.

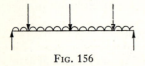

FIG. 156

However, before we progress to the general case of Fig. 156, which shows a beam supporting a number of so-called concentrated loads in addition to its own weight, we shall obtain our definitions for bending moment and shearing force from the simpler case of Fig. 157.

Here, in Fig. 157, we have a beam of negligible weight, simply supported at its ends and carrying a single concentrated load at the centre. This load and the two reactions are all acting in the plane of the paper.

Since the beam is in equilibrium the reactions will each be $\frac{W}{2}$ units.

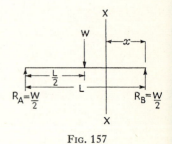

FIG. 157

For this state of equilibrium let us consider what is happening at any section XX of the beam, distance x from the right-hand end.

There will be a tendency for the beam to bend upwards or rotate about XX due to the right-hand reaction $\frac{W}{2}$, and at the same time this reaction will tend to shear the right-hand part of the beam upwards across XX.

By introducing equal and opposite forces of $\dfrac{W}{2}$ at XX, as in Fig. 158, the equilibrium of the beam is undisturbed, and it is then apparent that at XX there occurs simultaneously a COUPLE or BENDING MOMENT $M_1 = \dfrac{W}{2}x$, and a SHEARING FORCE $V_1 = \dfrac{W}{2}$.

(a) Actual loading   (b) Equivalent system

FIG. 158

For equilibrium this external bending moment and shearing force must be balanced by internal forces acting over the cross-section at XX. These are supplied by the action of the beam to the left of XX on that part of the beam to the right of the section. It is important to realise that there can only be one value for the bending moment and shearing force at any section of a beam. To establish this fact consider the beam to the left of XX : Fig. 159.

The two external forces W and $\dfrac{W}{2}$ give a resultant $V_2 = W - \dfrac{W}{2} = \dfrac{W}{2}$ acting at XX, together with a couple M_2 equal to the algebraic sum of the moments of W and $\dfrac{W}{2}$,

$$\text{i.e. } M_2 = \frac{W}{2}(L - x) - W\left(\frac{L}{2} - x\right),$$

$$\therefore \ M_2 = \frac{W}{2}x.$$

Hence we have shown that

$$V_1 = V_2 = \frac{W}{2} = V$$

$$M_1 = M_2 = \frac{W}{2}x = M.$$

The force V, which is equal to the algebraic sum of all the external forces acting on the beam to the left OR right of the section considered, is called the SHEARING FORCE at that section. The couple M, which is equal to the algebraic sum of the moments of all the external forces to the left OR right of the section considered, is called the BENDING MOMENT at that section.

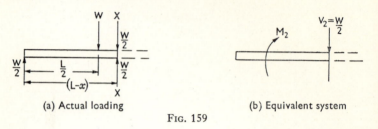

(a) Actual loading   (b) Equivalent system

FIG. 159

It should be appreciated that if M_1 represents the external bending moment at XX due to forces to the right of XX, then M_2 is equivalent to the internal 'moment of resistance' offered by the beam at XX. They are thus equal in magnitude and opposite in direction.

The internal forces represent the resistance to bending and shearing which the beam must offer if failure is not to occur. Consequently bending and shearing STRESSES are set up in the beam causing it to bend and deflect.

Summary

(i) There is only one value of bending moment and shearing force at a given section regardless of which side of the section is considered.

(ii) The Bending Moment at any section of a beam is the algebraic sum of the moments of the forces to one side of the section.

(iii) The Shearing Force at any section is the algebraic sum of the forces to one side of the section.

We are now able to determine values for B.M. and S.F. at a given section of beam, as shown in example 61.

Example 61

A beam is simply supported over a span of 20 feet and carries loads of 2, 3 and 4 tons at distances of 2, 6 and 14 feet respectively from the left-hand support. Determine the bending moment and shearing force at sections 5 feet and 10 feet from the left-hand support. Neglect the weight of the beam.

Reactions

Take Moments about R_A—Fig. 160.

C.W. Moments $=(2 \times 2)+(3 \times 6)+(4 \times 14)=$**78 ton ft.**
A.C.W. Moments $=R_B$ **20 ton ft.**

For Equilibrium

$20 . R_B = 78$ $R_B = $**3·9 tons** $R_A = 9 - 3·9 = $**5·1 tons.**

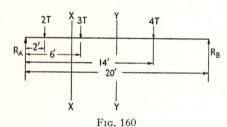

FIG. 160

Section XX at 5 feet from R_A

Consider forces to left of XX.

Net S.F. $= 5·1 - 2 = $**3·1 tons.**
B.M. $= (5·1 \times 5) - (2 \times 3) = $**19·5 ton ft.**

Section YY at 10 feet from R_A

Consider forces to left of YY.

Net S.F. $= 5·1 - 5 = $**0·1 ton.**
B.M. $= (5·1 \times 10) - (2 \times 8) - (3 \times 4) = $**23 ton ft.**

Our next step must be to take into account the weight of the beam or what is in effect the same thing, any additional uniformly distributed loading on the beam.

For the purpose of calculating reactions and bending moments only, a uniformly dis-

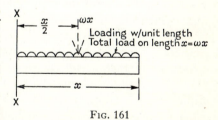

FIG. 161

tributed load may be replaced by an equivalent imaginary concentrated load acting, as shown dotted in Fig. 161. This substitution must not be used, of course, when calculating the shearing force at any section.

Example 62

A uniform cantilever of length 12 feet weighs 18 cwt. and carries an end load of 6 tons. Determine the bending moment and shearing force at the centre of the beam and at the fixed end.

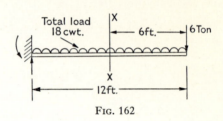

Fig. 162

Since the beam is uniform the load per foot run of beam $= 1\frac{1}{2}$ cwt. $= 0.075$ ton.

Section XX.—Fig. 162.

> S.F. at XX = Total load to right of XX
> $= (6 \times 0.075) + 6 = $ **6·45 tons.**
> B.M. at XX $= (6 \times 6) + (6 \times 0.075 \times 3) = $ **37·35 ton ft.**

Wall (fixed end)

> S.F. at Wall = Total load to right of Wall
> $= (12 \times 0.075) + 6 = $ **6·9 tons.**
> B.M. at Wall $= (6 \times 12) + (12 \times 0.075 \times 6) = $ **77·4 ton ft.**

54. Notation for Bending Moment and Shearing Force.— It is not enough that we should know only the numerical values for the bending moment and shearing force at any section. We need to know in addition HOW the beam reacts under load. Fig. 163, which illustrates the various loading arrangements with corresponding deflections, will indicate that we are concerned with two types of bending at any section—that which causes hogging and that which causes sagging of the beam. We can, therefore, readily indicate the type of bending which occurs by calling the bending moment either positive or negative. A similar convention can be adopted for shearing force by considering the direction in which shear takes place.

Although the choice of notation is purely arbitrary, it is important to state clearly the one which has been used. The

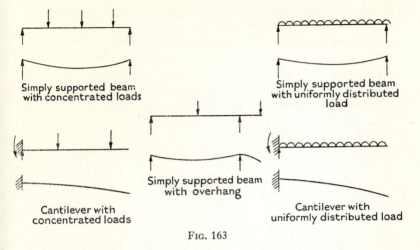

Simply supported beam with concentrated loads

Simply supported beam with uniformly distributed load

Cantilever with concentrated loads

Simply supported beam with overhang

Cantilever with uniformly distributed load

Fig. 163

notation adopted in this text is given in Fig. 164, and may be stated as follows :

Bending Moment : Sagging or Convex downwards bending negative.
Hogging or Convex upwards bending positive.

Shearing Force : If right-hand section of beam tends to move downwards relative to the left—negative shear.
If right-hand section tends to move upwards relative to the left—positive shear.

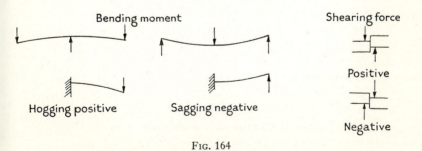

Bending moment

Shearing force

Hogging positive

Sagging negative

Positive

Negative

Fig. 164

55. Bending Moment and Shearing Force Diagrams.—
If the bending moment and shearing force is calculated at a
number of sections of the beam, and these values are plotted
vertically with the beam as horizontal axis, graphs of the variation
of bending moment and shearing force are obtained, normally
called bending moment and shearing force diagrams.

Before we begin plotting such figures it is necessary that one
point which often causes difficulty with the shearing force diagrams
is fully explained.

Consider the cantilever with a single load.   Fig. 165.

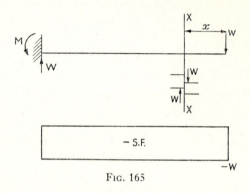

Fig. 165

The shearing force at section XX distance x from the free end
is $-W$.

Since this value is independent of x the shearing force must be
constant along the beam.

For two loads the shearing force diagram is shown in Fig.
166 (ii).   There is only one force to the right of XX.   Therefore
the shearing force at section XX is $-W_1$, and this value is only
true from C to B.

There are two forces to the right of YY.   Therefore, the
shearing force at Section YY is $-(W_1 + W_2)$, and this value is only
true from B to A.

From inspection of Fig. 166 (ii) there would, at first, appear
to be two possible values of shearing force at section B.   However
the student should appreciate that real point loads rarely exist
and that the actual conditions are illustrated by Fig. 166 (iv).
Here W_2 is more truly represented as uniformly distributed over a

length 'a'. The smaller the length 'a' is made the nearer the actual diagram becomes to the diagram (ii), which assumes point loading.

Since the shearing force under the loads is hardly ever required, this latter diagram is the one that is used in this branch of mechanics.

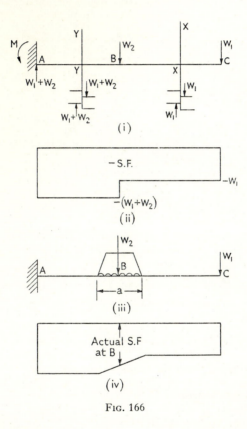

Fig. 166

56. Standard Cases.—The principles already developed will now be applied to the simple cases of beam loading given in Fig. 163.

The numerical examples following each case should indicate the method of solving the revision problems at the end of the chapter.

Case (i).   *Cantilever with Concentrated End Load.*   *Fig.* 167

Consider a section distance x from the free end.

The B.M. at section XX $= + Wx$.

This is the equation of a straight line, and, therefore, the maximum B.M. will occur when x is a maximum, i.e. when $x = L$, Max. B.M. $= + WL$.

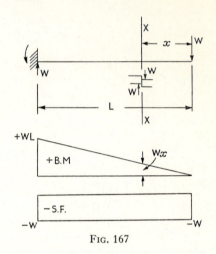

FIG. 167

The S.F. at section XX $= - W$.   This is independent of x and therefore, gives constant shearing force along the beam.

Example 63

Plot Bending Moment and Shearing Force diagrams for the Cantilever shown.   Fig. 168.

Neglect the weight of the beam.

Bending Moment Diagram

B.M. at A		ZERO.
B.M. at B $= +(1 \times 3)$		$=$**3 ton ft.**
B.M. at C $= +(1 \times 6)+(2 \times 3)$		$=$**12 ton ft.**
B.M. at D $= +(1 \times 10)+(2 \times 7)+(2 \times 4)$		$=$**32 ton ft.**

Shearing Force Diagram

S.F. A $-$ B	$= -$**1 ton.**
S.F. B $-$ C $= -1-2$	$= -$**3 tons.**
S.F. C $-$ D $= -1-2-2$	$= -$**5 tons.**

The dotted lines in the bending moment and shearing force diagrams illustrate that these diagrams are made up of diagrams for the separate loads added together.

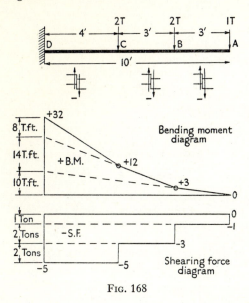

FIG. 168

Case (ii). Cantilever with Uniformly Distributed Load. Fig. 169

Consider a section distance x from free end.

Total load on a length $x = \omega x$.

B.M. at $XX = + \omega x \cdot \dfrac{x}{2}$
$$= + \dfrac{\omega x^2}{2}.$$

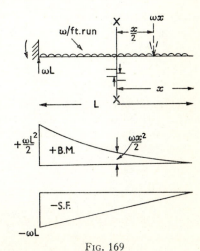

This is the equation of a parabola having a maximum value when $x = L$.

Max. B.M. $= \dfrac{\omega L^2}{2}.$

Since the total load on the beam $\omega L = W$, Max. B.M. $= \dfrac{WL}{2}.$

The S.F. at $XX = - \omega x.$

FIG. 169

This is the equation of a straight line giving maximum shearing force when x is a maximum, i.e. when $x = L$ at the wall.   Max. S.F. $= -\omega L$.

Example 64

A cantilever, 12 feet long, weighs 1 ton/ft. of its length and carries a load of 4 tons at 6 feet from the wall.   Fig. 170.

Draw B.M. and S.F. diagrams.

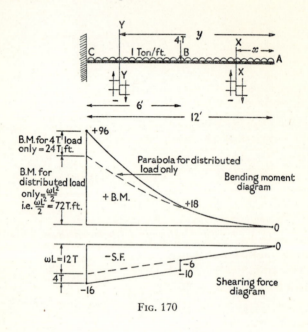

FIG. 170

Bending Moment Diagram

B.M. at A   ZERO.

$$\text{B.M. at XX} = +\left(\omega x \cdot \frac{x}{2}\right) = +\frac{\omega x^2}{2}$$

from $x = 0$
to $x = 6$.

$$\text{B.M. at B} = +\left(1 \times 6 \cdot \frac{6}{2}\right) = +\frac{36}{2} = +18 \text{ ton ft.}$$

when $x = 6$

B.M. at YY $= + \left(\omega y \cdot \dfrac{y}{2} \right) + 4(y - 6)$

from $y = 6$
to $\quad y = 12$.

B.M. at C $= + \left(1 \cdot \dfrac{144}{2} \right) + 4(12 - 6) = +96$ ton ft.

when $y = 12$.

Shearing Force Diagram

$$\text{S.F. A} - \text{B} \qquad = -(\omega x)$$
$$\text{S.F. at B when } x = 6 = -(1 \times 6) \qquad = -6 \text{ tons.}$$
$$\text{S.F. B} - \text{C} \qquad = -(\omega y) - 4$$
$$\text{S.F. at C when } y = 12 = -(1 \times 12) - 4 = -16 \text{ tons.}$$

Case (iii). *Simply Supported Beam with Central Concentrated Load. Fig. 171*

Consider a section XX distance x from the right-hand support. The B.M. at XX

$$= -\frac{W}{2} x. \qquad \qquad (1)$$

Equation (1) is true for values of x from O to $\dfrac{L}{2}$ ONLY.

The B.M. at YY

$$= -\frac{Wy}{2} + W\left(y - \frac{L}{2} \right)$$

$$= -\frac{WL}{2} + \frac{Wy}{2}. \qquad (2)$$

Equation (2) is true for values of y from $\dfrac{L}{2}$ to L ONLY.

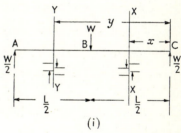

(i)

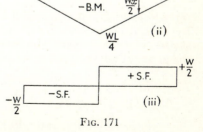

(ii)

(iii)

FIG. 171

These two equations enable the principal values of the bending moment to be obtained.

The B.M. at B is found by substituting

$$x = \frac{L}{2} \text{ in equation (1)}$$

or

$$y = \frac{L}{2} \text{ in equation (2).}$$

Hence B.M. at B

$$= -\frac{W}{2}\frac{L}{2} = -\frac{WL}{4} \text{ using equation (1)}$$

$$= -\frac{WL}{2} + \frac{W}{2}\frac{L}{2} = -\frac{WL}{4} \text{ using equation (2).}$$

The B.M. at C is found by substituting $x = O$ in equation (1), i.e. B.M. at $C = O$, and the B.M. at A by substituting $y = L$ in (2), i.e. B.M. at $A = O$.

Hence the Bending Moment diagram may be drawn at (ii).

The S.F. at XX $= +\frac{W}{2}$ and is constant from $x = O$ to $x = \frac{L}{2}$, or until another load is reached at B.

The S.F. at YY $= +\frac{W}{2} - W = -\frac{W}{2}$, and is constant from $y = \frac{L}{2}$ to $y = L$.

Alternatively we could have considered the algebraic sum of the forces to the left of YY which gives S.F. $= -\frac{W}{2}$ direct.

Since the shearing force has constant values between AB and BC, the S.F. diagram may be drawn at (iii).

Example 65

A beam, 10 feet long, is simply supported at its ends and carries three loads of 2 tons, 5 tons and 1 ton placed respectively 3, 7 and 9 feet from the left-hand support. Neglecting the weight of the beam, sketch the bending moment and shearing force diagrams and mark on the principal values.

Reactions. Fig. 172

Take Moments about A.

Anti-C.W. Moments $= 10 R_E$ ton ft.
C.W. Moments $= (2 \times 3) + (5 \times 7) + (1 \times 9) = 50$ ton ft.

$$R_E = 5 \text{ tons.} \qquad R_A = 3 \text{ tons.}$$

Bending Moment Diagram

	B.M. at A and E	ZERO.
	B.M. at D $= -(5 \times 1)$	$= -5$ ton ft.
	B.M. at C $= -(5 \times 3) + (1 \times 2)$	$= -13$ ton ft.
	B.M. at B $= -(5 \times 7) + (1 \times 6) + (5 \times 4)$	$= -9$ ton ft.

Shearing Force Diagram

	S.F. E-D	$= +5$ tons.
	S.F. D-C $= +5 - 1$	$= +4$ tons.
	S.F. C-B $= 5 - 1 - 5$	$= -1$ ton.
	S.F. B-A $= +5 - 1 - 5 - 2$	$= -3$ tons.

By now it should be appreciated that bending moment and shearing force diagrams give values of bending moment and shearing force for every section of the beam. For some problems this may not be necessary, and for many complicated loading systems it is sufficient to know at which section the bending moment is a maximum. Or it may be required to know only the magnitude of the change in bending moment between any two sections. This information can be obtained from the much simpler shearing force diagram without plotting the bending moment diagram.

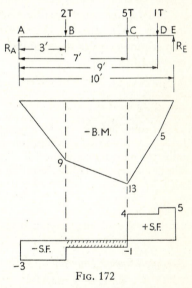

Fig. 172

What follows applies equally to ALL types of loading, but the example No. 65 is used as an illustration.

In Fig. 172 it will be observed that at C the bending moment is a maximum and that at this section the shearing force is zero or changes its sign from positive to negative shearing force. Therefore it is shown that :

'The bending moment is a maximum when the shearing force is zero or changes its sign.'

A.M.—P

Since there may be more than one such position along the beam the greatest of the bending moments at such points is the maximum bending moment (see example No. 69).

This fact has been deduced for a concentrated load system but is verified again by the calculus on page 220.

The shaded area of the S.F. diagram between B and C (Fig. 172) = 4 ton ft.

Since the bending moment at B is 9 ton ft. and that at C 13 ton ft., the change in bending moment between these two sections is seen to be 4 ton ft., and corresponds to the area of the shearing force diagram between B and C.

Consider two other sections such as A and E.

Change in bending moment between A and E   = zero.
Area of shearing force diagram between A and E

$$= +5 + 8 - 4 - 9 = \text{zero.}$$

No matter what two sections are chosen and whatever the loading, 'The area of the shearing force diagram between any two sections equals the change in bending moment between these two sections'.

Example 66

A beam, 20 feet long, simply supported at its ends, is symmetrically loaded by two loads of 2 tons placed 5 feet from each support. Determine the maximum values of B.M. and S.F. What important property can be deduced from the resulting diagrams ?

Solution

Since the loading is symmetrical, $R_A = R_D = 2$ **tons.**   Fig. 173.

Bending Moment Diagram

B.M. at A and D   ZERO.
B.M. at B $= -(R_A \times 5) = -(2 \times 5) = -10$ ton ft.
B.M. at C $= -(2 \times 15) + (2 \times 10) = -10$ ton ft.

Consider any section XX between B and C.

B.M. at XX $= -(2 \times x) + 2(x - 5) = -10$ ton ft.,

i.e. B.M. is constant from B to C.

Shearing Force Diagram

$$\text{S.F. A-B} \qquad\qquad = +2 \text{ tons.}$$
$$\text{S.F. B-C} = +2 - 2 \qquad = \text{ZERO.}$$
$$\text{S.F. C-D} = +2 - 2 - 2 = -2 \text{ tons.}$$

This is a very important example and the student should observe that

(*a*) where the bending moment diagram is constant the shearing force is zero ;

(*b*) the shearing force is zero at B and C giving two positions of maximum bending moment ;

(*c*) the area of the shearing force diagram between B and C is zero, and this equals the change in B.M. between B and C.

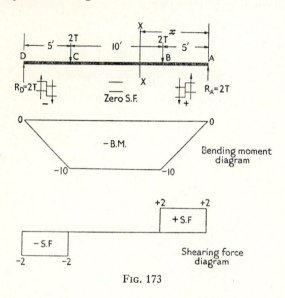

Fig. 173

Case (iv).   Simply Supported Beam with Uniformly Distributed Load.   Fig. 174

Consider a section distance *x* from the right-hand support. Fig. 174 (i).

The B.M. at XX

$$= -\frac{\omega L x}{2} + \frac{\omega x \cdot x}{2} \quad \text{or} \quad M_{XX} = -\frac{\omega L x}{2} + \frac{\omega x^2}{2} \qquad . \quad (1)$$

This is the equation of a parabola with its origin at the right-hand end of the beam.

To find where the B.M. is a maximum we know from the calculus that $\dfrac{dM}{dx} = O$ when M is a maximum.

Since

$$M = -\frac{\omega L x}{2} + \frac{\omega x^2}{2}$$

$$\frac{dM}{dx} = -\frac{\omega L}{2} + \omega x = O, \text{ i.e. } \frac{\omega L}{2} = \omega x,$$

∴ Maximum B.M. occurs when $x = \dfrac{L}{2}$.

Since the loading is symmetrical we could have expected this result.

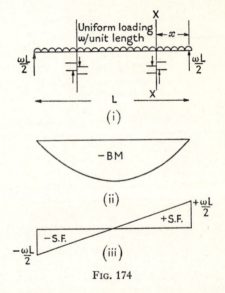

Fig. 174

Substituting $x = \dfrac{L}{2}$ in the equation for M_{xx} at (1),

$$\text{Maximum B.M.} = -\frac{\omega L}{2}\frac{L}{2} + \frac{\omega}{2}\left(\frac{L}{2}\right)^2 = -\frac{\omega L^2}{8}.$$

But the total load $W = \omega L$.   ∴ Maximum B.M. $= -\dfrac{WL}{8}$.

The S.F. at any section XX $= \dfrac{\omega L}{2} - \omega x$, which is the equation of a straight line.

The shearing force, therefore, varies with x and will have maximum values of opposite sign when $x = 0$ and when $x = L$.

Note.—The Bending moment is a maximum when the shearing force is zero.

57. Contraflexure or Inflexion.—So far we have dealt with beams loaded to give either sagging or hogging bending. Now we must consider the case of a beam with loading outside one of the supports.

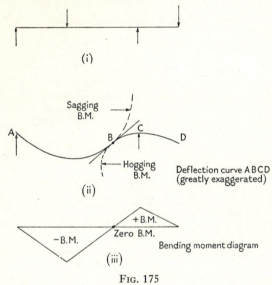

Fig. 175

It is easily seen that a beam loaded as in Fig. 175 combines bending conditions of opposite sign and that the deflected form changes in direction at some point B. This point—where the sign of the bending moment changes—is called either a point of contraflexure, i.e. opposite flexure, or a point of inflexion, i.e. a point at which bending and change of direction occurs. Both terms are in use and apply equally to the point B.

When a bending moment changes from positive to negative or vice versa, it must pass through zero, and, therefore, a point of contraflexure or inflexion is a point of zero bending moment.

The bending moment diagram for such a beam is illustrated in Fig. 175 (iii).

Example 67

The two supports of a 12 foot beam are placed 10 feet apart to give an overhang at the right-hand end of 2 feet. This overhang carries a distributed load of $\frac{1}{2}$ ton/ft. run. Additional concentrated loads of 2 tons and 1 ton are placed at 3 feet and 7 feet from the left-hand support respectively. Sketch S.F. and B.M. diagrams, and show on the B.M. diagram any points of contraflexure.

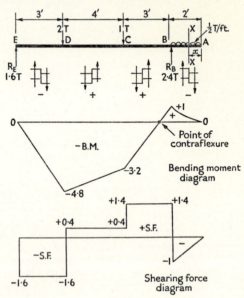

Fig. 176

Solution

Reactions

Take moments about E.   Fig. 176.

Anti-C.W. Moments $= 10\ R_B$ ton ft.
C.W. Moments $= (2 \times 3) + (1 \times 7) + (1 \times 11) = 24$ ton ft.

$$\therefore R_B = \textbf{2·4 tons.} \qquad R_E = \textbf{1·6 tons.}$$

Bending Moment Diagram

B.M. at A and E   ZERO.
B.M. at B $= +(1 \times 1)$ $= +1$ ton ft.
B.M. at C $= +(1 \times 4) - (2·4 \times 3)$ $= -3·2$ ton ft.
B.M. at D $= +(1 \times 8) + (1 \times 4) - (2·4 \times 7) = -4·8$ ton ft.

Shearing Force Diagram

S.F. A-B $= -\omega x$

S.F. at B when $x = 2 = -(\frac{1}{2} \times 2) = -1$ ton.

S.F. B-C $= -1 + 2 \cdot 4$ $\qquad = +1 \cdot 4$ tons.

S.F. C-D $= -1 + 2 \cdot 4 - 1$ $\qquad = +0 \cdot 4$ ton.

S.F. D-E $\qquad\qquad\qquad = -1 \cdot 6$ tons.

Example 68

A beam FA, 20 feet long, is simply supported at 'F' and 'B', 16 feet apart. The beam carries a uniformly distributed load of 1 ton/ft. run only between the supports, a vertical concentrated load of 6 tons

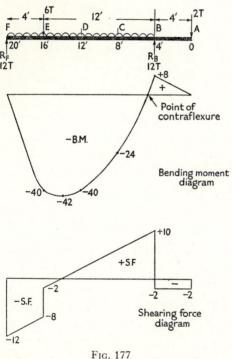

Fig. 177

at E, which is 4 feet from F, and a vertical concentrated load of 2 tons at A. Draw the S.F. and B.M. diagrams to scale and mark thereon the principal values and the position of the point of contraflexure on the beam. State the position and value of the maximum bending moment.

Solution No. 68.   Fig. 177

Reactions

Take moments about F.

A.C.W. Moments $= 16 \ R_B$ ton ft.
C.W. Moments $= (2 \times 20) + (16 \times 8) + (6 \times 4) = 192$ ton ft.

$$R_B = \textbf{12 tons.} \qquad R_F = \textbf{12 tons.}$$

Bending Moment Diagram

B.M. at A and F		ZERO.
B.M. at B $= + (2 \times 4)$		$= + 8$ ton ft.
B.M. at C $= + (2 \times 8) - (12 \times 4) + (4 \times 2)$		$= - 24$ ton ft.
B.M. at D $= + (2 \times 12) - (12 \times 8) + (8 \times 4)$		$= - 40$ ton ft.
B.M. at E $= + (2 \times 16) - (12 \times 12) + (12 \times 6)$		$= - 40$ ton ft.

Shearing Force Diagram

S.F. A-B		$= - 2$ tons.
S.F. just beyond B $= - 2 + 12$		$= + 10$ tons.
S.F. at D	$= - 2 + 12 - (1 \times 8)$	$= + 2$ tons.
S.F. at E	$= - 2 + 12 - (1 \times 12)$	$= - 2$ tons.
S.F. just beyond E $= - 2 + 12 - (1 \times 12) - 6$		$= - 8$ tons.

Further intermediate values can be calculated for each diagram.

The Maximum B.M. occurs where S.F. is zero. Maximum B.M. $= \textbf{42 ton ft.}$ and occurs $\textbf{6 ft.}$ from R_F.

Example 69

A uniform beam ABCD is 30 feet long and is simply supported at B and C, where AB $= 7$ feet and BC $= 20$ feet, AB and CD being overhangs. The beam ABCD carries throughout the whole of its length a uniformly distributed load of 0·5 ton per ft. run.

Calculate : (*a*) the position and value of the maximum bending moment on the beam ; (*b*) the positions of the points of contraflexure.

Sketch the Bending Moment and Shear Force diagrams and mark on all the principal values.

Solution No. 69.   Fig. 178

Reactions

Take moments about C.

A.C.W. Moments $= 20 R_B + (3 \times 0·5) 1·5$ ton ft.

$$\text{C.W. Moments} = (27 \times 0·5)(27 \times 0·5) = \frac{27^2}{4} \text{ ton ft.}$$

$$\therefore \ 20R_B = \frac{27^2}{4} - \frac{9}{4} = 180.$$

$$R_B = \textbf{9 tons.} \qquad R_C = \textbf{6 tons.}$$

For A-B. B.M. $= +\dfrac{\omega x^2}{2}$, for $x = 0$ to $x = 7$ ft.

$\therefore$ B.M. at B $= \dfrac{0.5}{2} \cdot 7^2 = \textbf{12·25 ton ft.}$

For D-C. B.M. $= +\dfrac{\omega y^2}{2}$, for $y = 0$ to $y = 3$ ft.

$\therefore$ B.M. at C $= \dfrac{0.5}{2} \cdot 3^2 = \textbf{2·25 ton ft.}$

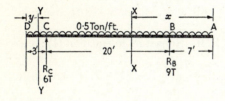

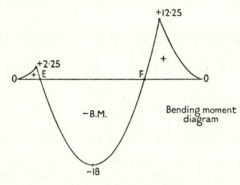

Bending moment diagram

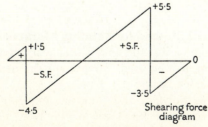

Shearing force diagram

Fig. 178

For B-C.   B.M. $= +\dfrac{\omega x^2}{2} - 9(x-7)$

$$\therefore \ M_{XX} \quad = \dfrac{x^2}{4} - 9(x-7). \qquad . \qquad . \qquad . \qquad (1)$$

To find Max. B.M. between B and C, differentiate equation (1) and equate to zero,

i.e. $\dfrac{dM_{XX}}{dx} = \dfrac{x}{2} - 9 = 0. \qquad \therefore \ x = 18$ ft. from A.

Alternatively

Max. B.M. occurs when S.F. is zero,

i.e. when $\omega x \downarrow \ = R_B \uparrow$ or $\dfrac{x}{2} = 9$ or $x = 18$ ft. from A.

Max. B.M. at $x = 18$ ft.   Substitute in equation (1):

$$\text{Max. B.M.} = \dfrac{(18)^2}{4} - 9(18-7) = -18 \text{ ton ft.}$$

There will be two points of contraflexure since the B.M. changes SIGN twice.

At points of contraflexure Equation (1) is zero,

i.e. $\dfrac{x^2}{4} - 9(x-7) = 0. \qquad \therefore \ \dfrac{x^2}{4} - 9x + 63 = 0.$

Solving for x gives $x = \dfrac{36 \pm 12\sqrt{2}}{2}.$

$\therefore \ x = 9\cdot516$ ft. (point F) and $x = 26\cdot484$ ft. (point E).

The question states 'sketch the B.M. and S.F. diagrams'. Hence one or two points on the diagram between B and C from equation (1) will be sufficient.

58. Graphical Method for Bending Moment Diagrams.—

It will now be shown that the link or funicular polygon dealt with in Chapter I can also be used to represent the bending moment diagram for a loaded beam:

Let the force polygon and funicular polygon be drawn to scale for the given beam as shown in Fig. 179. *de* in the force polygon gives reaction R_2 and *ea* the reaction R_1.

The bending moment at any section XX distance x from R_1 is numerically equal to $\{R_1 x - W_1(x - L_1)\}$, and we have now to show that this bending moment is represented to *scale* by the ordinate pg of the funicular polygon.

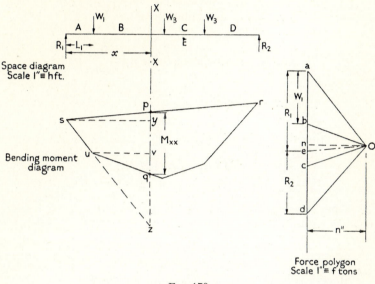

FIG. 179

The construction lines required on the funicular polygon are shown dotted from the funicular polygon.

$$pq = pz - qz.$$

Line sp on funicular polygon is parallel to eo on force polygon.

$$sz \text{ is parallel to } oa.$$
$$pz \text{ is parallel to } ea.$$

Hence triangles spz and oea are similar.

In the same way triangles syz and one are similar and uqz and oba are similar.

From these similar triangles it may be shown that

$$pz = \frac{sy \cdot ea}{on} \text{ and } qz = \frac{uv \cdot ab}{on}.$$

Then
$$pq = \frac{1}{on}(sy \cdot ea - uv \cdot ab). \qquad . \qquad . \qquad . \qquad (1)$$

Now
$$R_1 \text{ (ton)} = ea \text{ (inches)} \times \text{Scale } f\left(\frac{\text{ton}}{\text{inches}}\right),$$

$$\therefore ea = \frac{R_1}{f} \text{ and similarly } ab = \frac{W_1}{f}$$

$$x \text{ in ft.} = sy \text{ (inches)} \times \text{Scale } h\left(\frac{\text{ft.}}{\text{inches}}\right),$$

$$\therefore sy = \frac{x}{h} \text{ and } uv = \frac{x - L_1}{h}.$$

Substituting these values in (1) we get

$$pq = \frac{1}{on}\left\{\frac{x}{h}\frac{R_1}{f} - \left(\frac{x - L_1}{h}\right)\frac{W_1}{f_1}\right\}$$

or taking all the scale factors outside the bracket

$$pq = \frac{1}{on \cdot h \cdot f}\{R_1 x - W_1(x - L_1)\}.$$

Since BM is in ton ft., if pq is in inches, then

$$pq \times (on \cdot h \cdot f) = \{R_1 x - W_1(x - L_1)\} \text{ ton ft.,}$$

i.e. pq (inches) $\times on$ (inches) $\times f\left(\dfrac{\text{ton}}{\text{inches}}\right) \times h\left(\dfrac{\text{ft.}}{\text{inches}}\right) = M_{xx}$ ton ft.

This is the result obtained by taking moments about section XX, so that we have shown that the funicular polygon represents the bending moment diagram.

Example 70

A beam, 20 feet long, is supported at the left-hand end and at 4 feet from the right-hand end. It carries the following loads : 4 tons at 4 feet and 6 tons at 10 feet from the left-hand end, and 2 tons at the extreme right-hand end of the beam. Determine, graphically, the bending moment diagram and state the value of the reactions and the maximum bending moment.

Solution

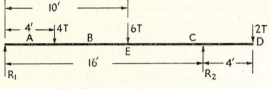

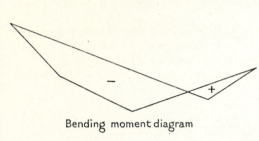

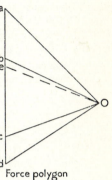

Bending moment diagram

Force polygon

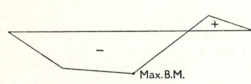

Max.B.M.

Bending moment diagram
redrawn on horizontal base

Reactions	$R_1 = ea = 4 \cdot 75$ Tons.
	$R_2 = de = 7 \cdot 25$ Tons.
Max. B.M.	$23 \cdot 5$ Ton ft. at 6 Ton load.

FIG. 180

EXERCISES VII

1. The diagram in Fig. 181 shows a ceiling beam which carries the concentrated loads indicated, these being due to the weight of machines secured to the floor above. Assuming the beam to be simply supported at its ends, draw the shear force and bending moment diagrams. All the calculations must be shown and the following scales employed :
 (Linear 1 in. = 10 ft.   S.F. 1 in. = 2 tons.   B.M. 1 in. = 20 ton ft.)
I.Prod.E.

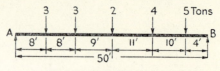

FIG. 181

2. A beam, 18 feet long, simply supported at one end and at another place 15 feet away so that there is an overhanging end, carries the loads shown in Fig. 182.

Carry out any necessary calculations and draw the shear force and bending moment diagrams. Use the scales :

Linear scale 1 in. =4 ft.   Shear scale 1 in. =10 tons.
Bending moment scale 1 in. =20 ton ft.   I.Prod.E.

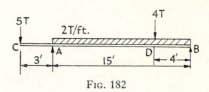

FIG. 182

3. Sketch the shear force and bending moment diagrams for the beam loaded as shown in Fig. 183.

Calculate beam reactions and significant values.   I.Prod.E.

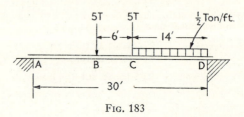

FIG. 183

4. Define the terms Bending Moment and Shearing Force.

Draw the Bending Moment and Shearing Force diagrams for the loaded beam shown in Fig. 184, stating the principal values on the diagrams.

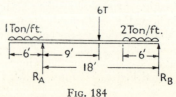

FIG. 184

5.   Define the terms Bending Moment and Shearing Force.

The beam in Fig. 185 is simply supported at A and B. Draw, for the loading given, the shearing force and bending moment diagrams, stating the sign convention.

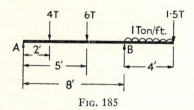

FIG. 185

6.   A beam is simply supported at points A and B and overhangs each end. If the arrangement and loading are as shown in Fig. 186, draw the shearing force and bending moment diagrams, giving numerical values at critical points. State where the minimum bending moment will occur between the supports, and determine its value.

N.C.T.E.C.

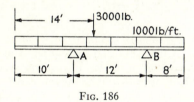

FIG. 186

7.   Draw the bending moment and shearing force diagrams for the cantilever shown in Fig. 187 and state the position of the point of contraflexure.

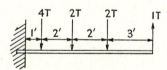

FIG. 187

8.   A horizontal beam is 40 feet long, has supports 25 feet apart, the overhangs being 5 feet and 10 feet at the left- and right-hand ends respectively. The beam supports vertical loads of 5 tons at the left and 7 tons at the right extremity, with a uniform distributed load of $\frac{1}{2}$ ton per foot run over the whole span. Draw to scale the bending moment and shearing force diagrams, indicating on the diagrams the principal values.

9. A horizontal beam is simply supported at its ends, which are 20 feet apart, and carries a uniformly distributed load of 500 lb./ft. run together with a concentrated load of 500 lb. at the middle of the span. Draw the bending moment and shearing force diagrams to scale.

10. A beam ABCDEF is simply supported at A—on the left—and at E, 24 feet from A. AB = 8 feet ; BC = 8 feet ; CD = 4 feet ; DE = 4 feet ; EF = 6 feet. There is a uniform load of 1 ton per foot run over the length AB and concentrated loads of 3, 2 and 4 tons at C, D and F respectively. Calculate the values of shearing force and bending moment at the lettered points, and show by sketches the forms of the shearing force and bending moment diagrams. Determine the position between the supports at which the shearing force is zero and evaluate the bending moment at that point. I.Mech.E.

11. A beam AB, 20 feet long, is freely supported at its ends A and B. It carries a uniformly spread load of 2 tons/ft. from A to the centre C, and also a concentrated load of 5 tons at D, 4 feet from B.

(a) Sketch the shear force and bending moment diagrams and insert on the diagrams the principal values.

(b) State the position and value of the maximum bending moment. U.E.I.

12. A beam, 60 feet long, is supported horizontally by vertical reactions at one end A, and at a point B, 40 feet from A. There is a uniformly distributed load of 0·5 ton/ft. over the whole beam, a concentrated load of 12 tons at the middle point of AB and another concentrated load of 8 tons at the end of the overhanging section. Draw the shearing force and bending moment diagrams, and calculate the position and magnitude of the maximum bending moment between A and B. N.C.T.E.C.

13. A horizontal beam ABCD, 25 feet long, is simply supported at A and C. The beam carries a uniformly distributed load of 0·25 ton/ft. between A and B and concentrated loads of 3 and 2 tons at B and D respectively.

(AB = 16 feet, AC = 20 feet.)

Sketch (not necessarily to scale) the shear force and bending moment diagrams, showing the values at the points A, B, C and D.

Determine the position and magnitude of the maximum bending moment on the beam. U.L.C.I.

14. A beam RS, of length 20 feet, rests on two supports, one at each end. A load of 10 cwt. is applied at a point 5 feet from R and another load of 15 cwt. 16 feet from R. The reactions are known to be 1781 lb. and 2363 lb. at R and S respectively. Calculate the position and magnitude of the load which keeps the beam in equilibrium. Draw the bending moment and shearing force diagrams for the fully loaded beam

to scales of cwt. ft. and cwt. respectively. What is the position and magnitude of the maximum bending moment ?

15. A beam AB, 18 feet long, is freely supported at B and at C, 6 feet from A. It carries a uniformly distributed load of 2 tons/ft. over its whole length AB, together with a concentrated load of 10 tons at D midway between C and B.

(*a*) Sketch the shear force and bending moment diagrams for the beam and insert on the diagrams the principal values.

(*b*) 'Calculate' the distance from A of the point of contraflexure.

U.E.I.

16. Give the meaning and significance of the term 'point of contra-flexure' as applied to beams.

ABCD is a beam of over-all length AD 18 feet, and is simply supported at B and C, where AB = 4 feet, CD = 2 feet, and BC = 12 feet. Concentrated loads of 4 tons and 8 tons act at A and D respectively. The span BC carries a uniformly distributed load of 2 tons/ft. run. Make dimensioned sketches of the shear force and bending moment diagrams. Determine the position of the points of contra-flexure.

17. A beam ABCD, 26 feet long, is simply supported at A and C, 22 feet apart. The beam carries a uniformly distributed load of 1 ton/ft. run only between the supports ; a vertical concentrated load of 3·5 tons at B, which is 4 feet from A, and a vertical concentrated load of 2 tons at D. Draw the S.F. and B.M. diagrams to scale and mark thereon the principal values and the position of the point of contra-flexure on the beam. Determine the position and value of the maximum bending moment.

18. A uniform steel girder, 20 feet long, and weighing 2000 lb., rests symmetrically on two supports. Determine the position of the supports so that the greatest bending moment shall be as small as possible, and for this position draw the shear force and bending moment diagrams. Insert on your diagrams the principal values and calculate the position of the points of contraflexure.

U.E.I.

19. A beam, 20 feet long, carries a load of 1 ton/ft. run uniformly distributed over its whole length. The beam is carried on two supports, each 4 feet from an end. Draw to scale the bending moment and shearing force diagrams. Find the maximum bending moment and the positions of the points of contraflexure. Determine also the position of the supports which will reduce the bending moment on the beam to a minimum.

20. Fig. 188 shows a simply supported beam with equal overhangs. The beam carries a concentrated load of 10 tons at the centre and three

A.M.—Q

uniformly distributed loads of ω ton/ft. run in the positions shown. If the maximum bending moment at any place on the beam is not to exceed 36·25 tons ft., find the value of ω and draw the bending moment and shearing force diagrams indicating principal values and the points of contraflexure.

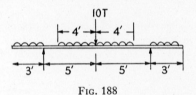

Fig. 188

Chapter VIII

BENDING OF BEAMS

59. Pure Bending.—We have already mentioned that when a beam is loaded internal forces occur within the beam. These opposing internal forces represent the resistance offered by the beam to bending and shearing and give rise to 'bending and shearing stresses'.

From Chapter VI we know that whenever stresses are set up in a component, distortion occurs, and in the case of a beam the bending and shearing stresses cause deflection of the beam to take place as indicated in Fig. 164 of the previous chapter.

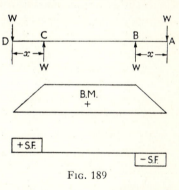

Fig. 189

For the general case of beam loading there is a shearing force and bending moment at each section of the beam, and consequently both stresses will occur simultaneously. Hence, both these types of stress will contribute towards the resulting deflection that takes place. Since, however, it is difficult to calculate the stresses set up when both shearing and bending occur together, we shall first consider the case of a beam in which the shearing force is zero.

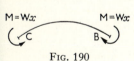

Fig. 190

For such a beam (Fig. 189) it is seen that between C and B the bending moment is constant and the shearing force zero. This case is called 'pure bending' for the length of the beam CB as shown in Fig. 190.

60. Bending Stresses.—Following the treatment of the previous chapter, if at any section XX the external bending moment on the beam is $M_1 = Wx$, then for equilibrium the internal stresses must provide an 'internal moment of resistance' or couple M_2 equal and opposite to M_1—Fig. 191.

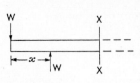

Our problem, then, is to determine how the stresses are distributed across XX in order to provide this internal moment M_2.

It is necessary, even for the simple treatment offered here, to make a number of assumptions which are given as follows : *

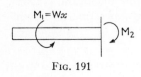

FIG. 191

(i) the beam is considered to be made up of an infinite number of longitudinal layers ;

(ii) each layer is in no way influenced by its adjacent layers but remains in contact with them during bending ;

(iii) each layer of fibres is subjected to longitudinal stress only ;

(iv) the value of Young's Modulus E is the same for compression as for tension and all the stresses are within the elastic limit,

(v) straight lines such as 'aa' and 'bb'—Fig. 192—marked on the beam before bending remain straight after bending—or, in other words, plane cross-sections of the beam before bending are plane after bending.

The assumptions (i)-(iii) imply that each fibre has the same radius of curvature when bent.

Consider a section of a rectangular beam as in Fig. 192.

Under the action of pure bending the two vertical lines aa and bb are rotated to the new positions a^1a^1 and b^1b^1, and will subtend some angle θ as shown. It is evident that the *separate* layers of fibres will be stressed—those near the top of the beam will be in tension and those near the bottom in compression. Thus it is not difficult to imagine that where the stress changes from tensile to compressive there will be one fibre layer represented by SS that is unstressed. This fibre layer is called the Neutral layer or Neutral Surface, and its intersection with any cross-section of the beam is called the NEUTRAL AXIS—NA as shown in Fig. 192.

Let us consider the strain induced by the bending of a fibre such as 'cc'—distance y above the Neutral Surface.

* See note on page 241.

Now the original length of cc = SS on the Neutral Surface.

Then if R be the radius of curvature of the Neutral Surface the original length of cc = Rθ.

This is so because SS remains unstressed and, therefore, unchanged in length.

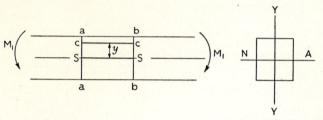

Plane of bending Y-Y

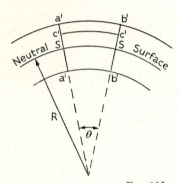

Fig. 192

New length of cc when bent $= (R + y)\theta$.

Therefore, change in length of cc $= (R + y)\theta - R\theta$.

But strain $= \dfrac{\text{Change in length}}{\text{Original length}}$.

Hence strain in cc $= \dfrac{(R + y)\theta - R\theta}{R\theta} = \dfrac{y}{R}$.

If f be the longitudinal stress in cc then since

$$\text{stress} = \text{Strain} \times E$$

$$f = \frac{y}{R} \times E, \qquad . \qquad . \qquad . \quad (1)$$

or $\dfrac{f}{y}$ is a constant *for a given curvature.*

Equation (1) clearly shows that the stress distribution across the beam is linear and that the stress on any layer is proportional to the distance of that layer from the Neutral Surface, as in Fig. 193. Hence the stress is zero when y is zero—that is, where the stress passes from compressive to tensile.

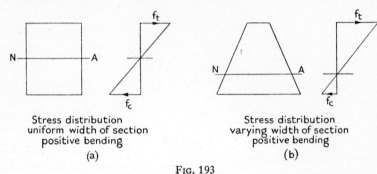

Stress distribution
uniform width of section
positive bending

(a)

Stress distribution
varying width of section
positive bending

(b)

FIG. 193

61. Neutral Axis.—Equation (1) verifies the existence of a neutral surface on which the stress is zero. It now remains for us to establish the position of this surface in relation to the beam section.

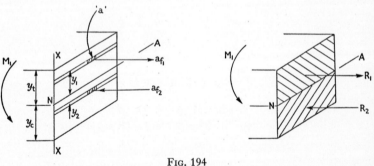

FIG. 194

If we consider an elemental strip of area 'a' on each side of the Neutral Surface where the stresses are, say, f_1 and f_2, then the force on these small areas will be af_1 and af_2 above and below the Neutral Surface respectively.

The resultants of all such internal forces will be

$$\sum_0^{y_t} af_1 = R_1 \text{ and } \sum_0^{y_0} af_2 = R_2,$$

as shown in Fig. 194.

But the external pure couple M_1 can only be balanced by an equal 'internal' pure couple provided by the resultant forces R_1 and R_2.

Hence, for R_1 and R_2 to form a pure couple and so provide for horizontal equilibrium, R_1 must equal R_2 in magnitude; then algebraically

$$R_1 + R_2 = 0.$$

This equation, based on Fig. 168 indicates that the net algebraic internal force on the section must be zero.

Therefore
$$\sum_0^{y_t} af_1 + \sum_0^{y_c} af_2 = 0,$$

and by substituting from equation (1) $f = \dfrac{E}{R}y$,

we obtain
$$\frac{E}{R}\sum_0^{y_t} ay_1 + \frac{E}{R}\sum_0^{y_c} ay_2 = 0,$$

or
$$\sum_0^{y_t} ay_1 + \sum_0^{y_c} ay_2 = 0.$$

But the left-hand side represents the total first moment of area of the section about the Neutral Axis,

$$\therefore \ A\bar{y} = 0,$$

where   A = Total area of section, and

   $\bar{y}$ = distance of centroid from the Neutral Axis.

Now if $A\bar{y} = 0$, A cannot equal zero.

Therefore $\bar{y}$ must equal zero, which means that the centroid coincides with the neutral axis.

This last statement is true regardless of the shape of the beam cross-section.

62. Internal Moment of Resistance.—The resisting force on an elemental area 'a' at distance y_1 above the Neutral Axis has been given as af_1.

From Fig. 194 the moment of this resisting force about the Neutral Axis = af_1y_1. But from Equation (1) $f_1 = \dfrac{E}{R}y_1$.

Therefore, by substitution, the Moment of the force on the elemental area

$$= a\left[\frac{E}{R}y_1\right]y_1$$

$$= a\frac{E}{R}y_1{}^2.$$

Hence, the sum of all such resisting moments above and below the Neutral Axis M_2

$$= \sum_{-y_c}^{+y_t}\frac{E}{R}ay_1{}^2,$$

or M_2

$$= \frac{E}{R}\sum_{-y_c}^{+y_t}ay_1{}^2,$$

where the expression

$$\sum_{-y_c}^{+y_t}ay_1{}^2$$

is called the second moment of area of the section about the Neutral Axis and is given the Symbol I.

Therefore $\qquad\qquad M_2 = \frac{EI}{R}.$ $\qquad$. $\qquad$. $\qquad$. $\quad$ (2)

In view of the similarity of the expression Σay^2 with Σmy^2 (see Chapter IV) the second moment of area is often very erroneously referred to as the moment of inertia.

Since, for equilibrium, the Moment of Resistance M_2 must equal the Applied Bending Moment M_1

$$M_2 = M_1 = M.$$

Therefore, combining equations (1) and (2), we get the simple bending relationship

$$\frac{M}{I} = \frac{E}{R} = \frac{f}{y}.$$

Units

For the student's reference a summary of the units and terms employed in bending problems will now be given.

I   The linear dimensions of the beam are usually given in inches.   Hence $I = \Sigma ay^2$ will have units (inches)$^2 \times$ (inches)$^2 = $ (inches)^4.

E and f Both these terms have units of stress, i.e. lb. per sq. inch or tons per sq. inch.

M   Bending Moments are usually given in lb. inches, or ton inches (consistent with beam dimensions in inches).

y The distance from the neutral axis at which the fibres are subjected to a stress, f. y, is given in inches.
Since the symbol y is also used to represent beam deflection care should be taken with its use in the bending equation.

EI   The product EI is called the Flexural Rigidity of the beam and has units lb. inches^2 or ton inches^2.

Z   When y represents the distance of the extreme fibres from the neutral axis the ratio of second moment of area to y is called the Modulus of the section Z,

$$\text{i.e. } Z = \frac{I}{y}.$$

Z has units $\dfrac{(\text{inches})^4}{\text{inches}} = (\text{inches})^3$.

This ratio enables the permissible bending moment on a beam to be obtained directly from the simple equation $M = fZ$.

In order to familiarise the student with the method of presenting data relating to beam sections, Fig. 195 shows extracts from List 6 dealing with beams in British Standard 4″ Dimensions and properties of British Standard Channels and Beams for structural purposes. Attention is drawn to the use of the letter J in this list to represent 'Moment of Inertia'.

Note on Simple Bending Theory

The theory just outlined provides a fairly accurate means of calculating bending stresses in beams subjected to pure bending only. However, as already pointed out, the general case of beam loading will involve shearing forces which may produce shearing stresses throughout the beam length. Fortunately, it is found by experiment that the simple theory is not limited to the case of pure bending but will give satisfactory results for varying bending moments—provided that stresses are not required too near the load points.

In view of this, further studies will indicate that it is usual to treat the general case in two steps. First, the bending stresses produced by the bending moment alone are calculated, according to the beam theory given and then the shearing stresses set up by the shearing forces are calculated separately.

Naturally the assumptions made for pure bending will not then apply but the theory has been found sufficiently accurate for most problems encountered in practical engineering.

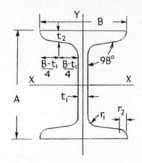

Size	Approximate Weight per foot	Standard Thickness		Radii		Sectional Area	Centre of Gravity		Moments of Inertia		Radii of Gyration		Modulus of Section
$A \times B$		Web. t_1	Flange t_2	Root r_1	Toe r_2	a	cx	cy	Jx	Jy	ix	iy	Zx
Inches	lb.	Inches		Inches		Inches²	Inches		Inches⁴		Inches		Inches
12×5	32	0·35	0·550	0·53	0·26	9·45	0	0	221·07	9·69	4·84	1·01	36·84
13×5	35	0·35	0·604	0·53	0·26	10·30	0	0	283·51	10·82	5·25	1·03	43·62
22×7	75	0·50	0·834	0·69	0·34	22·06	0	0	1676·80	41·76	8·72	1·36	152·44

The data shown above have been reproduced by permission of the British Standards Institution, 2 Park Street, London W.1, from whom official copies of the Standard may be purchased.

FIG. 195

Example 71

A circular section beam of 1 inch diameter and second moment of area 0·0492 inch⁴ is subjected to a pure bending couple of 0·75 ton inches. Determine the maximum stress set up in the material and also the radius of curvature of the beam $E = 30 \times 10^6$ lb. per sq. in.

Solution

The maximum stress will occur at the extreme fibres, i.e. where $y = 0·5$ inch.

Then
$$f_{max} = \frac{M}{I}y = \frac{0·75 \times 2240 \times 0·5}{0·0492}$$

$$f_{max} = 17100 \text{ lb. per sq. in.}$$

From $\dfrac{M}{I} = \dfrac{E}{R}$, we get $R = \dfrac{EI}{M}$, where R is the radius of Curvature.

Then
$$R = \dfrac{30 \times 10^6 \times 0 \cdot 0492}{0 \cdot 75 \times 2240}$$

R = 880 inches.

Example 72

A cantilever of uniform I section carries loads of 4 tons at 11 feet from the wall and 2 tons 8 feet from the wall. If the second moment of area of the section is 360 (inches)⁴, determine the necessary depth of section in order to limit the maximum tensile and compressive stresses to 5 tons/in.²

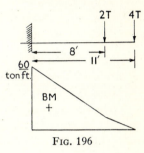

Fig. 196

Solution

From the expression $\dfrac{M}{Z} = f$ it is clear that the maximum tensile and compressive stresses will occur where the bending moment is a maximum.

Max. B.M. at wall $= (4 \times 11) + (2 \times 8) = $ **60 ton ft.**

Then from
$$\dfrac{M}{I} = \dfrac{f}{y}, \ y = \dfrac{fI}{M} = \dfrac{5 \times 360}{60 \times 12}$$
$$= 2 \cdot 5 \text{ inches.}$$

Since beam is uniform, depth of section $= 2y = $ **5 inches.**

63. Calculation of Second Moments of Area.—The expression Σay^2 has been called the second moment of area of a section and is given the symbol I.

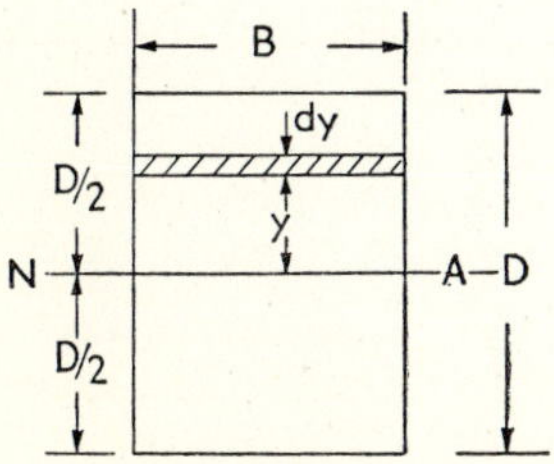

The value of I will now be obtained for typical beam sections.

(i) Rectangle. Fig. 197.

Consider an elemental strip thickness dy and distance y from the NA.

Fig. 197

Area of strip $= B dy$
Second Moment of area of strip about NA $= B dy \cdot y^2$

Total Second Moment for whole section $= \displaystyle\int_{-D/2}^{+D/2} B dy \cdot y^2$

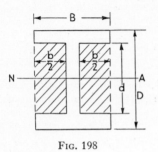

FIG. 198

$$I_{NA} = \frac{BD^3}{12}.$$

(ii) I Section. Fig. 198.

I_{NA} for this section is the difference in I values for two rectangles,

i.e. $= \dfrac{BD^3}{12} - \dfrac{bd^3}{12}.$

(iii) Circular Section.*

$$I_{NA} = \frac{\pi D^4}{64}.$$

In the case of unsymmetrical sections the position of the Neutral Axis has first to be calculated.

When this position is known it is often more convenient to find I_{NA} by considering the sum of the second moments of area of individual sections about an axis through the base of each section. Such values can readily be obtained by means of the 'Parallel Axis Theorem'.

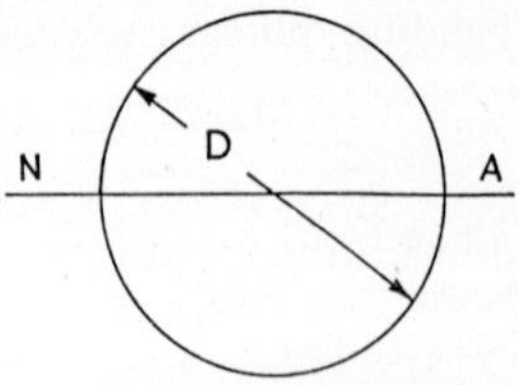

FIG. 199

The 'Parallel Axis Theorem' states that the second moment of *area* about 'any' axis is equal to the second moment of *area* about a parallel axis through the centroid plus the *area* of the section multiplied by the square of the distance separating the axes.

Consider the particular case of a rectangle and an axis XX coincident with its base. Fig. 200.

The theorem states that

$$I_{XX} = I_{NA} + A\left(\frac{D}{2}\right)^2$$

$$= \frac{BD^3}{12} + BD\,\frac{D^2}{4}$$

$$= \frac{BD^3}{3} = \text{Second Moment of Area about the base.}$$

* See Appendix for Perpendicular Axis Theorem.

This can easily be checked by integration about the base of the rectangle.

The above calculation clearly shows that I_{XX} is greater than I_{NA}. This is true for all sections and means that for all parallel axes the second moment of area will always be a minimum for the axis through the centroid.

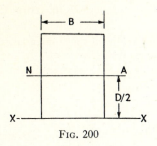

FIG. 200

Example 73

Determine the second moment of area about the neutral axis for the two sections shown in Fig. 201.

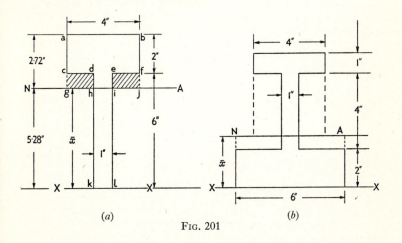

(a) (b)

FIG. 201

For both these sections the position of the NA must first be calculated. Since, for any axis, the first moment of area of the whole section will be equal to the sum of the first moments of the separate areas, the position of the NA can be found by taking moments about XX.

Solution (a)

Total Area of section $\qquad\qquad$ $= 14$ in.^{2}

Distance of Centroid of *abcf* from XX $= 6 + 1 = 7$ in.

Distance of Centroid of *dekl* from XX $= 3$ in.

Then $\qquad\qquad 14\bar{x} = (4 \times 2)7 + (6 \times 1)$

$$x = \frac{56 + 18}{14} = \textbf{5·28 in. from XX}$$

I_{NA} can now be calculated for all the individual rectangles which have their base coincident with the NA.

Above NA.

$$I_{NA} = Iabgj - (Icdgh + Iefij)$$

$$= \frac{4 \times (2·72)^3}{3} - \frac{3 \times (0·72)^3}{3} = \textbf{26·30 in.}^4$$

Below NA.

$$I_{NA} = Ihikl = \frac{1 \times (5·28)^3}{3} = \textbf{49·30 in.}^4$$

Total I_{NA} for whole section $= 26·30 + 49·30 = \textbf{75·6 in.}^4$

Solution (b)

Total Area of Section $= 20$ sq. in.

To find position of NA take Moments about XX :

$$20\bar{x} = (4 \times 1)6\tfrac{1}{2} + (4 \times 1)4 + (6 \times 2)1$$
$$\bar{x} = 2·70 \text{ in.}$$

$$\text{Total } I_{NA} = \frac{4 \times (4·3)^3}{3} - \frac{3 \times (3·3)^3}{3} + \frac{6 \times (2·7)^3}{3} - \frac{5 \times (0·7)^3}{3}$$

$$= 106 - 36 + 39·50 - 0·58.$$

$$I_{NA} = \textbf{108·9 in.}^4$$

Example 74

A cast-iron beam, 30 feet long, is simply supported at its ends and has a cross-section as shown in Fig. 201 (*b*). Given that the density of cast iron is 0·26 lb. per cubic inch, determine the maximum tensile and compressive stresses set up in the beam due to its own weight.

Solution

The second moment of area of the section was calculated in example 73 as **108·9 in.⁴**

For a beam simply supported at the ends and carrying a uniformly distributed load the maximum bending moment is given by :

$$M_{max} = \frac{WL}{8}, \text{ where } W = \text{Total Load on beam.}$$

Weight of beam $=$ Volume $\times$ Density. $\left[\text{in.}^3 \times \dfrac{\text{lb.}}{\text{in}^3} \right]$

$$= (20 \times 30 \times 12) \times 0\cdot26 \text{ lb.}$$

$$\therefore \; W = \textbf{1872 lb.}$$

Hence $\qquad M_{max} = \dfrac{1872 \times (30 \times 12)}{8}$, where L is in inches.

$$M_{max} = \textbf{84240 lb. in.}$$

Now for sagging bending moment the upper flange will be in compression—Fig. 202.

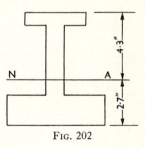

N ⎯⎯⎯⎯ A

FIG. 202

Maximum f_c will occur at extreme fibre where $y_c = 4\cdot3$ inches.

The lower flange will be in tension. Maximum f_t will occur at extreme fibre where $y_t = 2\cdot7$ inches.

Therefore $\qquad \dfrac{M}{I} = \dfrac{f_c}{y_c}$ and $f_c = \dfrac{84240 \times 4\cdot3}{108\cdot9}$

$$f_c = \textbf{3340 lb. per sq. in.}$$

and $\qquad \dfrac{M}{I} = \dfrac{f_t}{y_t} \quad \therefore \; f_t = \dfrac{84240 \times 2\cdot7}{108\cdot9}$

$$f_t = \textbf{2100 lb. per sq. in.}$$

For this example it will be instructive to investigate the effect of the web of the I section on the final value of the second moment of area of the whole section.

Total I_{NA} for whole section $= \textbf{108·9 in.⁴}$

I for the flanges alone can be found by finding the I about the centroid of each flange and transferring the axis as already indicated.

For Top Flange $I_{NA} = \dfrac{4 \times 1^3}{12} + 4 \times (3 \cdot 8)^2 = 58 \cdot 09$ in.⁴

For Bottom Flange $I_{NA} = \dfrac{6 \times 2^3}{12} + 12 \times (1 \cdot 7)^2 = 38 \cdot 68$ in.⁴

Total I_{NA} for both flanges $= 96 \cdot 77$ in.⁴

Therefore, of the total I for the section, the web only contributes

$$108 \cdot 9 - 96 \cdot 77 = 12 \cdot 13 \text{ in.}^4$$

Now if M is the resisting moment supplied by the whole section :

$$\% \text{ supplied by flanges} = \frac{96 \cdot 77}{108 \cdot 9} = 88 \cdot 8\%$$

$$\% \text{ supplied by web} = \frac{12 \cdot 13}{108 \cdot 9} = 11 \cdot 2\%.$$

This solution indicates that in cases where the web thickness is small relative to the other dimensions of the section it would be convenient as a first approximation to assume that the web contributes nothing to the resisting moment and just to consider the I value for the flanges alone.

Example 75

Fig. 203 shows a cantilever which carries a vertical load of 3400 lb. at the free end. Calculate the maximum tensile and compressive stresses in the material of the cantilever and show the stress distribution across the section.

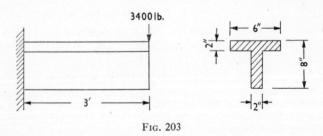

Fig. 203

Solution

To find the position of the Neutral Axis take moments about XX. Fig. 204.

Then $$24\bar{x} = (2 \times 6)7 + (2 \times 6)3$$

$$x = \frac{120}{24} = 5 \text{ in.}$$

Therefore $$y_c = 5 \text{ in.} \qquad y_t = 3 \text{ in.}$$

$$I_{NA} = \frac{6 \times 3^3}{3} - \frac{4 \times 1^3}{3} + \frac{2 \times 5^3}{3}$$

$$= 54 - \frac{4}{3} + 83\tfrac{1}{3}.$$

$$I_{NA} = 136 \text{ inches.}^4$$

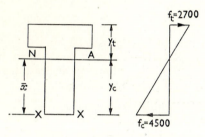

Fɪɢ. 204

From $$\frac{M}{I} = \frac{f}{y} \qquad f_c = \frac{M}{I}y_c \qquad f_t = \frac{M}{I}y_t$$

M = Max. Bending Moment $= 3400 \times 36 = 122400 \text{ lb. in.}$

Therefore $$f_c = \frac{122400 \times 5}{136} \text{ and } f_t = \frac{122400 \times 3}{136}$$

$$f_c = 4500 \text{ lb. per sq. in.} \qquad f_t = 2700 \text{ lb. per sq. in.}$$

Stress distribution as shown in Fig. 204.

Example 76

Fig. 205 illustrates a steel tube, 0·75 inch external diameter and bore 'd'. It is found that the central deflection at mid-span is 1 inch above the two supports.

Calculate (a) the bore 'd', (b) the maximum tensile stress in the tube. Young's Modulus of Elasticity $= 30 \times 10^6$ lb. per sq. in. Neglect the weight of the tube.

A.M.—R

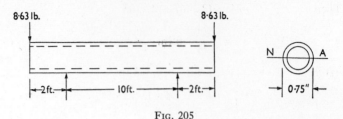

FIG. 205

Introduction

Since this question involves deflection the following note may prove helpful.

Consider two intersecting chords AD and BC in the circle of Fig. 206.

Then

Angle AOC = Angle BOD—opposite angles.
Angle CAD = Angle CBD—Triangles on the same base subtend the same angles at the circumference.
Angle ACO = Angle BDO—by subtraction.

Hence triangles AOC and BOD are similar,

$$\therefore \quad \frac{AO}{CO} = \frac{BO}{OD},$$

or $AO . OD = BO . OC.$

This gives the property of intersecting chords of a circle.

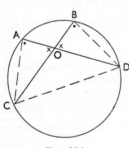

FIG. 206

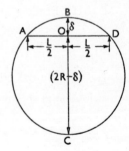

FIG. 207

Solution

The beam between the supports is subjected to a pure couple and will therefore bend in the arc of a circle as shown in Fig. 207.

Let deflection at centre $= \delta$.

From the above property of chords of a circle

$$\frac{L}{2} \cdot \frac{L}{2} = \delta(2R - \delta),$$

$$\therefore \frac{L^2}{4} = 2R\delta - \delta^2.$$

Now δ is small compared with R if stresses are within the elastic limit.
Therefore δ^2 can be neglected as being very small compared with $2R\delta$.

Hence $$R = \frac{L^2}{8\delta} \text{ will be used.}$$

Since the bore 'd' is unknown it will be necessary to evaluate I_N from $\dfrac{M}{I} = \dfrac{E}{R}$, using $R = \dfrac{L^2}{8\delta}$.

Transposing and rearranging

$$I = \frac{ML^2}{8E\delta},$$

M = Maximum bending moment = (8.63×24) lb. in.

$$\therefore I = \frac{(8.63 \times 24)(120)^2}{8 \times 30 \times 10^6 \times 1}$$

I = 0·01243 in.⁴

For a hollow circular beam $I = \dfrac{\pi}{4}[D^4 - d^4]$,

$$\therefore 0.01243 = \frac{\pi}{4}[(0.75)^4 - d^4]$$

from which $$d \simeq 0.50 \text{ in.}$$

Hence Max. stress $f = \dfrac{M}{I}y = \dfrac{(8.63 \times 24) \times 0.375}{0.01243}$

$$f = 6245 \text{ lb. per sq. in.}$$

Example 77

A beam, 20 feet long, is freely supported at the ends and carries a uniformly distributed load of 1 ton per foot run over the whole length of the beam, together with concentrated loads of 2 tons at 5 feet from each end. The beam is built up of standard I section, 12 inches deep, and having a second moment of area of 488 in.⁴ with a 12-inch wide by 0·75-inch thick plate riveted to each flange. Calculate the maximum bending stress in the beam.

Calculate also the percentage increase in the strength of the beam owing to the addition of the plates.

Solution

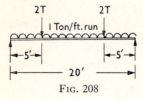

Fig. 208

Maximum bending moment due to distributed and concentrated loads occurs at centre of beam. Fig. 208.

$$M = (12 \times 10) - (2 \times 5) - (10 \times 5)$$

M = 60 ton ft. = 720 ton in.

Second Moment of Area of Joist = 488 in.⁴

$$\text{Second Moment of Area of Plates} = \left(\frac{12 \times 13\frac{1}{2}^3}{12}\right) - \left(\frac{12 \times 12^3}{12}\right) = 732 \text{ in.}^4$$

$$\text{Total I of Section} = 488 + 732 = \textbf{1220 in.}^4$$

$$f = \frac{M}{I} y \quad \therefore \quad f = \frac{720 \times 6\frac{3}{4}}{1220} = \textbf{3·98 tons per sq. in.}$$

Consider $\dfrac{I}{y}$ equals the Section Modulus. Then $f = \dfrac{M}{Z}$.

This equation indicates that the strength of a beam is directly proportional to its section modulus, i.e $M = fZ$.

Now   Z for joist and plates combined $= \dfrac{1220}{6\frac{3}{4}} = \textbf{180·7 in.}^4$

$$\text{Z for joist alone} = \frac{488}{6} = \textbf{81·3 in.}^4$$

Therefore   % increase in strength $= \left(\dfrac{180·7 - 81·3}{81·3}\right) \times 100$

$$= \textbf{122.3\%}$$

EXERCISES VIII

1. Prove the relationship between the stress and the radius of curvature of a beam subjected to simple bending. State the assumptions made. Determine the radius of curvature to which a steel strip, 0·2 inch thick and 1·5 inch wide, may be bent if the greatest stress is to be 6 tons/in.²

(Young's modulus for steel is 13,000 tons/in.²)

If a 2-feet length of steel strip is simply supported at its ends, find the central load which it can carry without the stress due to bending exceeding 6 tons/in.²

2. State the assumptions made in the theory of simple bending and establish the usual formulae for the calculation of stress and radius of curvature.

A steel strip, 2 inches broad and 0·1 inch thick, is bent round a wheel of 10 feet diameter. Calculate the bending moment necessary and the maximum stress set up in the strip.

(Young's modulus $= 30 \times 10^6$ lb./in.²)   I.Mech.E.

3. What two assumptions are made in the theory of elastic bending ?

An axle, 7 inches diameter and 7 feet 6 inches long, is supported in symmetrically placed bearings 6 feet apart. There is an upward load of 14 tons at each end of the axle. Draw the shearing force and bending moment diagram for the axle. Determine the maximum bending stress in the axle and the radius of curvature.

(Young's modulus for steel $= 30 \times 10^6$ lb./in.²)

N.C.T.E.C.

4. Determine the maximum bending moment which a circular bar of 2 inches diameter can carry, if the maximum bending stress is not to exceed 10 tons/in.² Establish from first principles any theory of bending formula you employ.   U.E.I.

5. A length of steel bar of rectangular section is bent to an arc of a circle of 50 feet radius. The width of the section is twice the depth and the depth is radial. If the ultimate strength of the steel is 30·6 tons/in.² and the modulus of elasticity of the steel is 30×10^6 lb./in.², determine the dimensions of the cross-section of the bar, using a factor of safety of 6. Also calculate the bending moment (in pounds feet) applied to the bar in the bent form.   U.L.C.I.

6. A steel beam, 12 feet long, rests symmetrically on supports 8 feet apart, and carries equal loads, W, at each end. The beam is of I section having flanges 5 inches wide and $\frac{1}{2}$ inch thick, a web $\frac{3}{8}$ inch thick and an overall depth of 12 inches.

(*a*) If the maximum allowable longitudinal stress in the beam is 9 tons/in.², determine, in tons, the greatest permissible value of W.

(*b*) For this value of W, calculate the height of the beam above the supports at the centre of the span.

Assume $E = 13,500$ tons/in.²   U.E.I.

7. Distinguish between the terms 'bending moment' and 'moment of resistance'. Write down the fundamental equations obtained from the simple theory of bending, explaining each term.

The hand lever of a bench shearing machine is 30 inches long. The link connecting the blade and the lever has pins $\frac{5}{8}$ inch diameter. When a force of 90 lb. is applied to the end of the lever, calculate the necessary thickness of the lever if the depth of the cross-section at the link is 2 inches.

Take the ultimate bending strength of the material as 60 tons/in² and allow for a factor of safety of 12. I.Prod.E.

8. (a) Derive the formula $f/y = M/I = E/R$ and state clearly the meaning of each of the symbols.

(b) A timber beam is 12 feet long and is freely supported at each end. It is required to carry over its entire span a distributed load of 4 tons. The depth of the beam is to be three times its breadth and the maximum longitudinal stress in the timber is not to exceed 1500 lb./in.² Determine the dimensions of the cross-section. U.E.I.

9. Explain carefully the meaning of the term 'second moment of area' and state the units in which it may be measured. A steel pipe, 4 inches external diameter and $\frac{1}{2}$ inch thick, is used as a beam to span a distance of 8 feet.

(a) Calculate the greatest central point load the pipe can carry if the longitudinal stress in the material is not to exceed 8 tons/in.²

(b) Compare the strength of this beam with that of a solid rod of the same cross-sectional area. U.E.I.

10. An R.S.J. has top and bottom flanges 6 inches × $\frac{3}{4}$ inch and a web $10\frac{1}{2}$ inches × $\frac{5}{8}$ inch, and is simply supported over a span of 10 feet. It carries a uniformly spread load of 1 ton/ft. What concentrated load may be carried in addition at the centre of the span, if the maximum stress due to bending must not exceed 5 tons/in.² ?

11. A girder, 20 feet long, is supported at each end and loaded with 4 tons and 6 tons at distances of 5 feet and 15 feet respectively from the left-hand end. Determine the maximum bending moment acting on the girder and the maximum stress induced.

The girder is of I section, 5 inches deep, width of flanges $4\frac{1}{2}$ inches, thickness of each flange 0·5 inch and web thickness 0·3 inch.

What is the radius of curvature of the girder at the section where the bending moment is a maximum ?

$$(E = 30 \times 10^6 \text{ lb./in.}^2)$$

12. A steel beam, 12 feet long, is of Tee section and has a flange 6 inches wide and an over-all depth of 6 inches. The flange and the web are both $\frac{1}{2}$ inch thick. The beam is freely supported at the ends with the flange uppermost and carries a point load at a distance of 4 feet from one support.

Calculate the greatest value of this load, if the maximum allowable tensile stress must not exceed 9 tons/in.² What is then the greatest compressive stress in the beam ? Ignore the weight of the beam.

U.E.I.

13. A cantilever has a free length of 8 feet. It is of T section with the flange 4 inches × 0·75 inch and web 8 inches × 0·5 inch. The flange is in tension. What load per foot run can be applied if the maximum tensile stress is limited to 2 tons/in.² ? What is the maximum compressive stress ?

14. A rolled steel joist of I section is 8 inches deep over-all and has equal flanges 4 inches wide and 1·0 inch thick. The web is 0·5 inch thick and centrally placed. The joist is horizontal and simply supported at the ends with the flanges horizontal over a span of 10 feet. The beam carries a uniformly distributed load of 0·25 ton/ft. over the whole span and a load of 1 ton concentrated at mid-span. Determine the maximum bending moment applied to the joist by the loading and the maximum stress produced due to this bending moment.

U.L.C.I.

15. Fig. 209 shows a cantilever carrying a uniformly distributed load of ω lb./ft. run. Determine the value of ω if the maximum bending stress in the material is 5000 lb./in.²

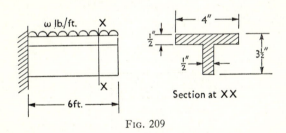

Section at X X

Fig. 209

16. Fig. 210 shows a trolley pole, 20 feet high, made from a steel tube 8 inches outside diameter, ½ inch thick. If the horizontal pull at the top of the pole is equal to 500 lb., find the maximum stresses in the pole due to bending. Sketch the stress distribution across the section at ground level and determine the maximum bending stress at the inside diameter of the pole.

N.C.T.E.C.

17. The T section shown in Fig. 211 is used as a beam with the 4 inches flange underneath. Determine the position of the neutral axis and the second moment of area of the section about this axis.

If the maximum allowable bending stress in the material is 6 tons/in.^{2} and the T bar is placed on supports, 8 feet apart, what is the greatest uniform load in tons per foot that can be placed on the beam between the supports ? Neglect the weight of the beam. Show on a diagram the variation of stress over the section due to bending.

N.C.T.E.C.

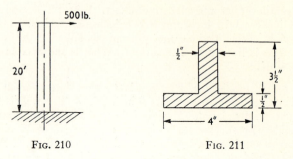

FIG. 210   FIG. 211

18. A cast-iron beam, 20 feet long, is simply supported at its ends and has a cross-section as shown in Fig. 212. Given that the density of cast iron is 0·26 lb./in.^{3}, determine the maximum tensile and compressive stresses set up in the beam due to its own weight.

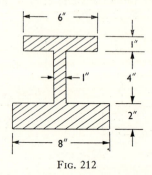

FIG. 212

19. A floor is to be carried on timber beams of rectangular section, pitched 15 inches apart, and simply supported on a span of 14 feet. The load on the floor is 100 lb./ft.^{2}, but, in addition, there is a concentrated load of 400 lb. per beam acting at a position 4 feet from one support. The maximum bending stress in the beams is not to exceed 800 lb./in.^{2} and the section is to have a depth three times the width. Sketch the shearing force and bending moment diagrams for the case. Evaluate the position and magnitude of the maximum bending moment and determine suitable beam dimensions. I.Mech.E.

20. A timber beam of rectangular section 9 inches broad by 12 inches deep is simply supported on a span of 24 feet when a uniformly distributed load of 250 lb./ft. is applied. The beam deflects excessively and a central support is introduced. This is fitted in such a way as to provide a central supporting reaction of 2750 lb. Draw to scale the shearing force diagrams for both cases. Calculate the maximum bending stresses before and after the change, stating where these occur. Show by a sketch the form of the bending moment diagram for the case with the central prop. I.Mech.E.

21. A beam simply supported at the ends of a span of 20 feet has to carry a uniformly distributed load of 1 ton/ft. run over the whole length and a central concentrated load of 8 tons. The beam is formed of a standard steel I section, 14 inches deep by 6 inches flange width, with a steel plate 8 inches broad by $\frac{5}{8}$ inch thick riveted to each flange. The moment of inertia of the I section alone is 533 inches⁴. The effect of rivet holes is neglected. Determine the maximum bending stresses in the plates and in the flanges and estimate the proportion of the total bending moment taken by the plates. I.Mech.E.

22. A steel beam of symmetrical section has flanges 8 inches broad by 1 inch thick and a web 10 inches deep by $\frac{1}{2}$ inch thick. The moment of inertia of this section about a neutral axis through the centre and parallel to the flanges may be taken as 527 inches⁴. In order to suit certain attachments equal amounts are machined off the edges of the top flange making the width 6·5 inches. Determine the change in the position of the neutral axis and in the value of the moment of inertia.

The beam, so modified, is simply supported on a span of 20 feet and loaded with a uniformly distributed load of 1 ton/ft. over the whole length and two concentrated loads each of 2 tons symmetrically placed 10 feet apart. Sketch the shearing force diagram for the case. Evaluate the maximum bending moment and the maximum stress in the top flange. I.Mech.E.

23. A steel bar of T section is to be used as a simply supported beam over a span of 10 feet. The flange of the bar is 5 inches in width and the total depth of the section is 6 inches ; the metal is 0·75 inch thick throughout. The beam has to carry three equal concentrated loads of W lb. each, one centrally placed and the others at the quarter points of the span. The bar is set with the flange on top, and the maximum tensile stress due to bending is not to exceed 7·5 tons/in.² Determine the allowable value for W. I.Mech.E.

Chapter IX

TORSION

64. Torsional Stress.—When a shaft is acted upon by a torque about its longitudinal axis it can be shown by experiment that provided the angle of twist is small :

 (i) all circular sections of the shaft remain sensibly circular during twist and their diameters are unchanged ;
(ii) plane cross-sections remain plane. (cf. bending theory).

Let us imagine the shaft in Fig. 213 to be made up of an infinite number of short cylinders. When a torque is applied to the shaft there will be relative movement between these cylinders as a result of the resisting shearing stresses set up between them. These shearing stresses will act in a tangential direction and will

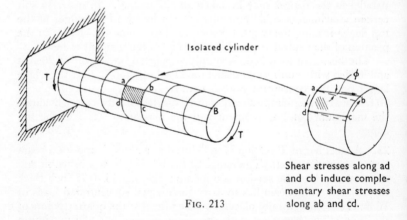

Isolated cylinder

Shear stresses along ad and cb induce complementary shear stresses along ab and cd.

Fig. 213

introduce complementary shear stresses as illustrated by the isolated disc in the diagram (see Chapter VI, page 166). Since the ends of the short cylinders remain parallel, any original rectangular element of material *abcd*, on the surface of the shaft, is in a state of pure shear and the effect of this stress distribution is to cause distortion of all the fibres in the longitudinal direction.

258

The lengths ab and dc of the element are each turned through some angle ϕ. As each individual cylinder will suffer the same distortion, the total effect will be that any original line AB drawn on the surface of the shaft parallel with the axis will take up the new position AB¹ when the torque is applied, i.e. it will move through the angle ϕ. (AB will actually form a helix when the shaft is twisted.)

Now for the case of pure shear the angle ϕ represents the shear strain.

Then if f is the shear stress at the surface of the shaft we have the relationship

$$\frac{\text{Shear Stress}}{\text{Shear Strain}} = \frac{f}{\phi} = G \text{ (Modulus of Rigidity)}$$

or
$$\phi = f/G. \qquad\qquad (1)$$

From Fig. 214 Arc BB¹ = $l\phi$ approx.
and also   Arc BB¹ = $r\theta$,

where θ is the angle of twist over the length of the shaft l.

$$\therefore\ l\phi = r\theta$$

and
$$\phi = \frac{r\theta}{l}.$$

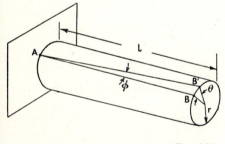
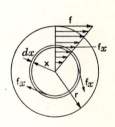

Fig. 214

Substituting this value of ϕ in equation (1) we get

$$\frac{r\theta}{l} = \frac{f}{G}$$

or
$$\frac{f}{r} = \frac{G\theta}{l}. \qquad\qquad (2)$$

From the experimental evidence given above θ is constant for all radii at a particular cross-section of the shaft. Therefore, since G and l are also constant it follows that equation (2) may be written

$$\frac{f}{r} = \text{Constant.} \qquad . \qquad . \qquad . \qquad (3)$$

Also if f_1 is the stress at any radius r_1, then $\dfrac{f_1}{r_1}$ is also a constant and, therefore, it follows that $\dfrac{f}{r} = \dfrac{f_1}{r_2} = \dfrac{f_2}{r_2}$, etc. = a constant.

This shows that the shear stress at any point within the cross-section of the shaft is proportional to the radius and that the stress increases uniformly from zero at the shaft centre to a maximum at the outside radius—Fig. 214.

65. Moment of Resistance.—Consider the cross-section of the shaft, Fig. 214, and let f_x be the stress acting over an elemental ring of fibres thickness dx at a radius x.

The shearing force on the elemental ring = Stress × Area of the ring,

$$\text{i.e. shearing force} = f_x \cdot 2\pi x \cdot dx.$$

But from equation (3) $\dfrac{f_x}{x} = \dfrac{f}{r}$,

$$\therefore f_x = \frac{f}{r} \cdot x.$$

Then shearing force $= \dfrac{f}{r} \cdot x \cdot 2\pi x dx.$

The moment of this shearing force about the shaft axis

$$= \frac{f}{r} \cdot x \cdot 2\pi x dx \cdot x$$

$$= \frac{f}{r} 2\pi x^3 dx.$$

The total internal resisting moment offered by the shaft material will be the sum of all such moments which, for equilibrium, must balance the applied torque T.

Hence $\displaystyle T = \int_0^{d/2} \frac{f}{r} 2\pi x^3 dx,$

and
$$T = \frac{f}{r}\int_0^{d/2} 2\pi x^3 dx.$$

If the expression
$$\int_0^{d/2} 2\pi x^3 dx$$

is replaced by the symbol J we get

$$T = \frac{f}{r}J, \text{ which, by rearrangement,}$$

becomes
$$\frac{T}{J} = \frac{f}{r}. \qquad \qquad (4)$$

J is called the 'polar second moment of area' of the cross-section about the shaft axis. In this equation f is the stress at the outside radius of the shaft and is a fixed value for a given applied torque. Combining this last equation (4) with equation (2) above we obtain the general torsion equation

$$\frac{T}{J} = \frac{f}{r} = \frac{G\theta}{l}.$$

If equation (4) is written in the form $T = f \cdot \frac{J}{r}$, then $\frac{J}{r}$ may be termed the polar modulus of section Z.

Hence
$$T = fZ.$$

This should be compared with the bending equation $M = fZ$, where the modulus of section Z in this case is equal to $\frac{I}{y}$. (See Chapter VIII, article 62.)

66. Polar Second Moment of Area.—For a solid shaft the summation of all the individual moments is extended from the shaft centre $x = o$ to the outside radius $x = \frac{d}{2}$, where $d =$ the shaft diameter.

Then
$$J = \int_0^{d/2} 2\pi x^3 dx$$
$$= \left[2\pi \frac{x^4}{4}\right]_0^{d/2}$$

or
$$J = \frac{\pi d^4}{32} \text{ (cf. } I = \frac{\pi d^4}{64} \text{ for bending).}$$

For a hollow shaft outside diameter D and inside diameter d the summation must obviously be taken over the limits $x = \dfrac{d}{2}$ to $x = \dfrac{D}{2}$,

$$\text{i.e. } J = \int_{d/2}^{D/2} 2\pi x^3 dx$$

or

$$J = \frac{\pi}{32}[D^4 - d^4].$$

Units

A summary of the units employed is given as follows :

T   The applied torque or internal resisting moment T has units lb. in., lb. ft., ton in., ton ft.

J   Polar Second Moment of Area J is usually given in inches^4 (consistent with shaft dimensions in inches).

G and f The modulus of rigidity and stress both have units lb./in.^{2} or ton/in.^{2}.

r The radius r from the shaft centre at which the stress is f given in inches.

θ Angle of twist θ in a length l inches is given in radians.

GJ   The product GJ is called the torsional rigidity of the shaft and has units lb. inches^2, ton inches^2 (cf. Flexural Rigidity EI for a beam).

67. Power Transmission.—When a shaft is used to transmit power it twists until the internal resisting moment balances the torque to be transmitted. If this torque is constant, then the angle of twist remains constant as outlined above and the work done per revolution will equal the torque multiplied by the angle turned through in one revolution.

$$\text{Work Done/rev.} = \text{Torque (lb. ft.)} \times 2\pi \text{ (radians)}$$
$$= T \times 2\pi \text{ ft. lb.}$$

If the shaft makes N revolutions per minute then

$$\text{Work Done/minute} = T \times 2\pi \times N \text{ ft. lb.}$$

$$\text{and Horse Power } = \frac{2\pi NT}{33000}.$$

In many cases of power transmission the torque is not constant but varies with time. This is illustrated by Fig. 215, which represents, diagrammatically, the torque diagram for a single-acting four-stage gas engine drawn on an angle base.

Clearly the horse power which can be transmitted in this case must be calculated on a *mean* or *average* torque value for the

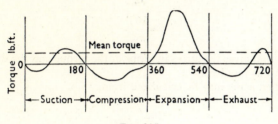

FIG. 215

whole cycle, but since the maximum stress condition must always be the criterion in shaft design it is the *peak* or *highest* value of the torque which is used when calculating the necessary shaft diameter. Hence, when dealing with problems which involve a variation in twisting moment, the above equations must be used in the form

$$\frac{T_{max}}{J} = \frac{f}{R} = \frac{G\theta}{l} \text{ and } HP = \frac{2\pi N T_{mean}}{33000}$$

Example 78

(a) Show that the shear stress induced in a shaft of circular cross-section by an axial torque is proportional to the distance from the axis of the shaft, and deduce an expression for this stress in terms of the torque and the dimension of the shaft.

(b) Calculate the diameter of a solid steel shaft to transmit 100 horse power at a speed of 240 rev./min., if the angle of twist on a length of 12 feet is not to exceed 0·4 degree. What is the maximum shear stress in the material ?

$G = 12 \times 10^6$ lb./in.² (U.E.I.)

Solution

(a) From equation (4) on page 261 we have $\frac{T}{J} = \frac{f}{r}$.

By rearrangement $f = \frac{T}{J} r$.

Since $J = \dfrac{\pi d^4}{32}$ and $r = \dfrac{d}{2}$, where $d = $ dia. of shaft,

$$f = \dfrac{T}{\dfrac{\pi d^4}{32}} \cdot \dfrac{d}{2}$$

or
$$f = \dfrac{16T}{\pi d^3}.$$

This equation gives the stress in terms of the torque and the dimension of the shaft as required.

(b) The torque may be calculated from :

$$HP = \dfrac{2\pi NT}{33000}.$$

Note.—this gives T in lb. ft. To convert to lb. in. multiply T by 12.

$$\therefore \quad 100 = \dfrac{2\pi \cdot 240 \cdot T}{33000}.$$

Hence $T = 2190$ lb. ft. or $T = 26,280$ lb. in.

Using this value of T in the equation

$$J = \dfrac{Tl}{G\theta},$$

where $\theta = $ angle of twist in radians ; $J = $ polar 2nd Moment of Area ; $l = $ length in inches ; $T = $ Torque in lb. in.

we get
$$\dfrac{\pi d^4}{32} = \dfrac{26280 \times 12 \cdot 12}{12 \times 10^6 \left(0 \cdot 4 \times \dfrac{\pi}{180}\right)}.$$

This gives $d^4 = 460$ in.^{4} or $d = 4 \cdot 62$ in.

To determine the maximum stress f we have

$$\dfrac{T}{J} = \dfrac{f}{r}.$$

Then
$$f = \dfrac{Tr}{J} = \dfrac{26280 \cdot r}{\dfrac{\pi}{32} d^4}.$$

Using $d^4 = 460$ in.^{4} and $r = 2 \cdot 31$ in.

$$f = \dfrac{26280 \times 2 \cdot 31}{\dfrac{\pi}{32} \cdot 460}. \quad f = 1350 \text{ lb. in.}^2$$

68. Hollow Shafts.—From the stress distribution shown in Fig. 214 it is seen that when the skin stress on the outside of the solid shaft reaches the safe limiting design value the material close to the axis is only subjected to a very low stress—in fact it becomes zero at the centre. For a hollow shaft the relationship $\frac{f}{r}$ = constant still holds, but for this case the stress now varies from the safe stress at the outside radius to a smaller value at the inside which, however, is not zero. For the same limiting stress it is, therefore, evident that the average stress intensity in a hollow shaft is greater than that in a solid shaft, and consequently for a given cross-sectional area a hollow shaft can transmit a greater torque.

This may be illustrated by comparing hollow and solid shafts of equal cross-sectional areas as follows :

If d and D are the inside and outside diameters of the hollow shaft and D_s the diameter of the solid shaft, we get for equal cross-sectional areas,

$$\frac{\pi}{4}D_s^2 = \frac{\pi}{4}(D^2 - d^2) \text{ or } D_s^2 = D^2 - d^2$$

and if $\qquad d = KD$, where K is a constant,

then $\qquad D_s^2 = D^2 - K^2D^2$

or $\qquad D_s = D\sqrt{1 - K^2}.$ . . . . (1)

Also from the torque equations for hollow and solid shafts having the same skin stresses

$$\left(\frac{Tr}{J}\right)_{\text{Solid}} = \left(\frac{Tr}{J}\right)_{\text{Hollow}}$$

$$\frac{T_s D_s/2}{\frac{\pi D_s^4}{32}} = \frac{T_H D/2}{\frac{\pi}{32}(D^4 - K^4D^4)}$$

and $\qquad \dfrac{T_H}{T_s} = \dfrac{D^3(1 - K^4)}{D_s^3}$

Substituting for D_s from equation (1) we get,

$$\frac{T_H}{T_s} = \sqrt{\frac{(1 + K^2)^2}{(1 - K^2)}}. \quad \text{Since } (1 - K^4) = (1 + K^2)(1 - K^2).$$

The advantage of the hollow shaft becomes apparent if we consider as an example the case when K = 0·8.

For this value of K the torque which can be transmitted by

A.M.—S

the hollow shaft is 2·73 times greater than is possible with a solid shaft of the same cross-sectional area. It is, therefore, evident that for a given speed of rotation a hollow shaft will transmit more power than a solid shaft of the same cross-sectional area whatever the diameter ratio.

Example 79

Determine the maximum torque that can be transmitted by a hollow shaft which has an external diameter of 10 inches and an internal diameter of 6 inches, if the working shear stress is limited to 4 tons/in.^{2} Also calculate the horse power that the shaft can transmit when running at 120 rev./min., if the maximum torque exceeds the mean by 25 per cent.

Solution

The maximum torque is associated with the maximum shear stress which occurs at the outside radius of the shaft.

Hence using
$$\frac{T_{max}}{J} = \frac{f}{r}$$

$$T_{max} = \frac{f}{\frac{D}{2}} \cdot \frac{\pi}{32}(D^4 - d^4), \text{ where } D = 10 \text{ in.}, d = 6 \text{ in.}$$

$$= \frac{4}{5} \cdot \frac{\pi}{32}(10^2 + 6^2)(10^2 - 6^2)$$

$$\therefore T_{max} = 684 \text{ ton in.}$$

Then
$$T_{max} = \frac{684 \times 2240}{12} \text{ lb. ft.}$$

$$T_{max} = \mathbf{127800 \text{ lb. ft.}}$$

Now
$$T_{max} = 1·25 \, T_{mean}$$

$$\therefore T_{mean} = \frac{127800}{1·25}$$

$$\therefore T_{mean} = 102200 \text{ lb. ft.}$$

Then
$$\text{Horse power} = \frac{2\pi N T_{mean}}{33000}$$

$$= \frac{2\pi \cdot 120 \cdot 102200}{33000}$$

$$\mathbf{H.P. = 2340.}$$

Example 80

A hollow shaft, with diameter ratio 3/5, is required to transmit 600 h.p. at 120 rev./min. with a uniform twisting moment. The shearing stress in the shaft must not exceed 4 tons/in.^{2} and the twist in a length of 8 feet must not exceed 1°. Calculate the minimum external diameter of the shaft satisfying these conditions. $G = 12 \times 10^6$ lb./in.^{2}

(U.L)

Solution

In this problem two conditions are laid down and the shaft diameter must be selected in order to satisfy them both.

From the given horse power and speed the torque must first be calculated from

$$\text{H.P.} = \frac{2\pi NT}{33000}$$

$$\therefore \ 600 = \frac{2\pi . 120 . T}{33000}.$$

Then $\qquad\qquad\qquad$ T $= 26300$ lb. ft. or T $= 315000$ lb. in.

The polar second moment of area can be expressed in terms of D, the outside diameter, by substituting $d = 3/5 D$ in the equation

$$J = \frac{\pi}{32}(D^4 - d^4)$$

$$\therefore \ J = \frac{\pi}{32}D^4\left(1 - \left(\frac{3}{5}\right)^4\right)$$

or $\qquad\qquad\qquad$ J $= 0.0855 D^4$.

From the torsion equation $\dfrac{T}{J} = \dfrac{f}{r} = \dfrac{G\theta}{l}$, $\dfrac{T}{J}$ can now be equated to $\dfrac{f}{r}$ to satisfy the maximum stress condition, and also equated to $\dfrac{G\theta}{l}$ for the maximum angle of twist condition.

<table>
<tr><td>Maximum Stress Condition</td><td>Maximum Angle of Twist Condition</td></tr>
<tr><td>$\dfrac{T}{J} = \dfrac{f}{r}$</td><td>$\dfrac{T}{J} = \dfrac{G\theta}{l}$</td></tr>
<tr><td>$\therefore \ f = \dfrac{T}{J} . \dfrac{D}{2}$</td><td>$\therefore \ \theta = \dfrac{Tl}{JG}$</td></tr>
</table>

$$f = \frac{315000}{0.0855D^4} \cdot \frac{D}{2} \qquad\qquad \theta = \frac{315000 \cdot 8 \times 12}{0.0855D^4 \cdot 12 \times 10^6}.$$

$$\therefore\ f = \frac{1840000}{D^3} \ . \qquad . \qquad . \quad (1) \qquad \therefore\ \theta = \frac{29.5}{D^4}. \qquad . \qquad . \quad (2)$$

But $\qquad f = 4 \times 2240$ lb. in.^{2} $\qquad$ and $\qquad \theta = 1 \times \dfrac{\pi}{180}$ radians

$\therefore\ D^3 = 206$ in.^{3} $\qquad\qquad\qquad \therefore\ D^4 = 1690$ in.^{4}

$\qquad\qquad$ **D = 5·90 in.** $\qquad\qquad\qquad\qquad$ **D = 6·41 in.**

Now although a shaft diameter of 5·9 inches satisfies the maximum stress condition, the substitution of this diameter in equation (2) above will clearly result in an angle of twist greater than 1°, which is not permissible. On the other hand, if a diameter of 6·41 inches (which satisfies the angle of twist condition) is substituted in equation (1), the resulting stress figure will be well within the allowable 4 tons/in.^{2}

Hence in order to satisfy both conditions the minimum diameter of shaft must be 6·41 inches.

Example 81

A hollow steel shaft, with an internal diameter 0·8 of the external diameter, is to be designed to transmit 75 horse power at 300 rev./min. The maximum shear stress is not to exceed 12,400 lb./in.^{2}, and the maximum twisting moment may be assumed to be 1·4 times the mean twisting moment. Calculate the required shaft diameters.

Two lengths of the shaft are to be connected by a flanged coupling, which is to have 4 bolts on a pitch circle diameter of 5 inches. Determine the necessary bolt diameter for a shear stress of 10,000 lb./in.^{2} in the bolts.

Solution

$$\text{H.P.} = \frac{2\pi N T_{mean}}{33000}$$

$$\therefore\ T_{mean} = \frac{75 \times 33000}{2\pi \cdot 300}$$

$$T_{mean} = 1315 \text{ lb. ft. or } T_{mean} = 15800 \text{ lb. in.}$$

Then $\qquad\qquad T_{max} = 1·4 \times 15800$ or $T_{max} = 22100$ lb. in.

$$J = \frac{\pi}{32}(D^4 - d^4)$$

or $$J = \frac{\pi}{32}D^4(1 - (0.8)^4), \text{ since } d = 0.8 \text{ D}.$$

Then $$J = \frac{\pi}{32}D^4 \cdot 0.59.$$

Substituting these values in the equation :

$$\frac{T_{max}}{J} = \frac{f}{D/2},$$

we get $$22100 = 12400 \cdot \frac{\pi D^3 \times 0.59}{16}$$

$$D^3 = 15.4$$

$$\therefore \text{ D} = 2.49'', \text{ say } \mathbf{D} = \mathbf{2.5 \text{ inches}}$$

Then $$d = 0.8 \times 2.5 \quad \mathbf{d} = \mathbf{2.0 \text{ inches}}$$

Now assuming that the torque carried by the bolts in the flanges must equal the torque on the shaft :

$$\text{Torque carried by each bolt} = \frac{22100}{4} = 5525 \text{ lb. in.}$$

Then the force on each bolt

$$= \frac{\text{Torque}}{\text{Pitch Circle Radius}} = \frac{5525}{2.5} = 2210 \text{ lb.}$$

Assuming the shear stress to be uniformly distributed over the cross-sectional area of each bolt :

$$\text{Stress} = \frac{\text{Force}}{\text{Area}}.$$

If bolt diameter $= d$. Then, $$\frac{\pi d^2}{4} = \frac{2210}{10000}$$

and $$d^2 = 0.281$$

$$\therefore \mathbf{d} = \mathbf{0.53 \text{ in.}}$$

No mention is made of a factor of safety, but it is likely that the bolt diameter used would be at least $\frac{5}{8}$ inch.

69. Composite Shafts.—Although problems involving shafts made up of different materials may be a little beyond the scope of some third-year syllabuses, the two problems chosen here illustrate that the solutions only require the application of principles already presented in the text.

It should be appreciated that where a composite shaft is made up of separate shafts connected in series the total applied torque will be transmitted by each shaft in turn, whereas if shafts are rigidly connected in parallel the total torque will be the sum of the torques transmitted by each individual shaft in the system—Fig. 216.

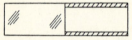

Hollow and solid shafts in series

Hollow and solid shafts in parallel

Fig. 216

Example 82

A solid alloy shaft of 2 inches diameter is to be coupled in series with a hollow steel shaft of the same external diameter. Find the internal diameter of the steel shaft, if the angle of twist per unit length is to be 75 per cent of that of the alloy shaft.

Determine the speed at which the shafts are to be driven to transmit 250 horse power, if the limits of shearing stress are to be 3·5 and 5 tons sq. in. in the alloy and steel respectively. The modulus of rigidity for steel $= 2·2 \times$ Modulus of rigidity for alloy.

(U. L.)

Solution

As this is a series problem the alloy shaft and the steel shaft transmit the *same* torque.

For the alloy shaft ;

$$\frac{T}{J_A} = \frac{G_A \theta_A}{l_A}$$

$$\therefore \ T = \frac{G_A \theta_A J_A}{l_A}.$$

Similarly for steel shaft :
$$T = \frac{G_s \theta_s J_s}{l_s}.$$

Equating these expressions for T, we get :

$$\frac{G_A \theta_A J_A}{l_A} = \frac{G_s \theta_s J_s}{l_s}.$$

But since $G_s = 2·2 G_A$ and $\dfrac{\theta_s}{l_s} = \dfrac{3}{4} \dfrac{\theta_A}{l_A}$,

we have
$$\frac{G_A \theta_A J_A}{l_A} = 2·2 G_A \frac{3}{4} \frac{\theta_A}{l_A} J_s,$$

from which $\qquad J_A = 1\cdot65\ J_S$

$$\therefore \ \frac{\pi \cdot 2^4}{32} = 1\cdot65\frac{\pi}{32}(2^4 - d^4),$$

where d is the i.d. of steel shaft.

and $\qquad\qquad\qquad d^4 = 6\cdot3$

$$d = 1\cdot59 \text{ in.}$$

For the alloy shaft the stress is limited to $3\cdot5$ tons/in.[2]

Hence equation $\dfrac{T}{J_A} = \dfrac{f_A}{r}$ can be used to calculate the torque associated with this maximum stress figure

$$T = \frac{3\cdot5 \times 2240}{1} \cdot \frac{\pi \cdot 2^4}{32} = 12350 \text{ lb. in.}$$

For the steel shaft the maximum stress is 5 tons/in.[2]

Therefore the torque which may be transmitted safely by this shaft is given by :

$$T = \frac{5 \times 2240}{1} \cdot \frac{\pi}{32}(2^4 - 6\cdot3) = 10700 \text{ lb. in.}$$

Any torque in excess of 10,700 lb. in. would cause a higher stress in the steel than is permitted. Hence the greatest permissible torque for the composite shaft is 10,700 lb. in., and the alloy shaft will not be working at its maximum allowable stress.

$$\text{H.P.} = \frac{2\pi NT}{33000}$$

or $$\qquad 250 = \frac{2\pi \cdot N \cdot 10700}{33000 \times 12}.$$

$$N = 1470 \text{ rev./min.}$$

Example 83

A steel rod is surrounded by a close-fitting tube of duralumin, the two being securely fastened together to form a composite shaft. Find the diameter of the steel rod and the outside diameter of the duralumin tube so that the maximum shearing stresses in the two materials do not exceed 6 and 4 tons/sq. in. respectively, when the composite shaft is subjected to a torque of $2\cdot75$ tons/in. Also calculate the angle of twist on a length of 4 feet. Modulus of rigidity for duralumin 1700 tons/in.[2] and for steel 5100 tons/in.[2]

(U. L.)

Solution

As the two shafts are rigidly fixed in parallel they will have the same angle of twist per unit length,

i.e. from the equation $\dfrac{f}{r} = \dfrac{G\theta}{l}$

we get : $\dfrac{\theta}{l} = \left(\dfrac{f}{rG}\right)_{\text{steel}} = \left(\dfrac{f}{rG}\right)_{\text{dura}}$

or $\dfrac{f_s}{\dfrac{d}{2}G_s} = \dfrac{f_D}{\dfrac{D}{2}G_D}$,

where d = o.d. of steel = i.d. of duralumin. D = o.d. of duralumin ; f_s and f_D are the maximum stresses in the steel and duralumin respectively.

$$\therefore \quad \frac{6}{\dfrac{d}{2}.5100} = \frac{4}{\dfrac{D}{2}.1700} \quad \text{or } \mathbf{D} = 2d.$$

Now for shafts in parallel :

Total torque = torque taken by duralumin + torque taken by steel.

$$\therefore \quad 2\cdot75 = \frac{f_D J_D}{D/2} \qquad\qquad + \frac{f_s J_s}{d/2}$$

or $\quad 2\cdot75 = \dfrac{4}{D/2} \cdot \dfrac{\pi}{32}(D^4 - d^4) \qquad + \dfrac{6}{d/2} \cdot \dfrac{\pi d^4}{32}$

and substituting $\quad D = 2d$

we get $\quad 2\cdot75 = 4 . \dfrac{\pi}{32} . d^3 . 15 + \dfrac{12 . \pi d^3}{32}$

$$\therefore \quad d^3 = 0\cdot388.$$

Hence : $\quad\quad\quad \mathbf{d = 0\cdot73 \text{ in. and } D = 1\cdot46 \text{ in.}}$

For the angle of twist :

$$\theta = \left(\frac{fl}{\dfrac{d}{2} . G}\right)_{\text{stee}}$$

or $\quad\quad \theta = \dfrac{6 \times 4 \times 12}{\dfrac{0\cdot73}{2} \times 5100} = 0\cdot155 \text{ radians.}$

$$\therefore \quad \theta = \mathbf{8\cdot9°}.$$

70. Close Coiled Helical Spring.—A helical spring is considered close coiled if the plane containing each coil is nearly perpendicular to the axis of the helix. This will be the case when the helix angle is small, and generally a spring is taken as close coiled if this angle is less than 10°.

The effect of an axial load on a close-coiled spring may be investigated by cutting the wire and considering the equilibrium of the upper part, as in Fig. 217.

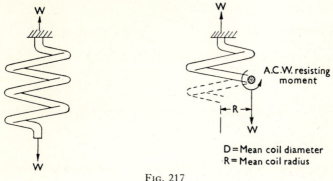

Fig. 217

It will be observed that for equilibrium the material of the wire must provide an internal resisting couple equal in magnitude and opposite in direction to the clockwise couple WR.

The couple WR is twisting the wire and sets up shear stresses within the material—in other words, the wire behaves like a shaft in torsion and the torsion equation will apply.

It should be appreciated that there will be a direct shear stress equal to the shearing force W divided by the cross-sectional area of the wire, but this will be small in comparison with the torsional stress and may be ignored. Likewise, any bending stress due to the obliquity of the coils is neglected in close-coiled spring calculations.

In the torsion formula θ will represent the total angle of twist over the length of the wire under the action of the torque WR. The length of the wire may be assumed to equal $2\pi RN$ where N is the number of effective coils, $2\pi R$ being the mean length of each coil.

From Fig. 217, if the axial load W is deflected an amount δ, then $\delta = R\theta$ approximately.

Spring stiffness S is defined as load per unit deflection, i.e. $S = \dfrac{W}{\delta}$.

In applying the torsion formula $\dfrac{T}{J} = \dfrac{f}{r} = \dfrac{G\theta}{l}$, it must be remembered that r is the outside radius of the wire where the maximum stress f occurs and is not R the mean radius of the coils.

Two useful spring formulae which express the deflection and maximum shear stress in terms of the spring dimensions and the loading may be derived as follows :

From the equation $\dfrac{T}{J} = \dfrac{G\theta}{l}$ we get by rearrangement $\theta = \dfrac{lT}{JG}$, since $T = WR$ and $l = 2\pi RN$,

then
$$\theta = \frac{WR \cdot 2\pi RN}{JG},$$

which may be substituted in the deflection equation $\delta = R\theta$, to give
$$\delta = \frac{WR^3 2\pi N}{JG}.$$

Replacing J by $\dfrac{\pi d^4}{32}$, where $d =$ the wire diameter, we get finally that deflection
$$\delta = \frac{64\,WR^3N}{Gd^4}.$$

Also from $\dfrac{T}{J} = \dfrac{f}{r}$, $f = \dfrac{WR \cdot r}{J}$.

Replacing R by D/2, r by $d/2$ and substituting for J, we have
$$f = \frac{W}{\dfrac{\pi d^4}{32}} \cdot \frac{D}{2} \cdot \frac{d}{2}$$

$$f = \frac{8\,WD}{\pi d^3}.$$

Example 84

A spiral spring is made from round steel closely coiled in 10 turns, whose outside diameter is 5 times that of the rod. If the spring is to stretch 0·5 inches under an axial load of 300 lb., find the diameter of the rod.

$G = 12,000,000$ lb. per sq. inch.

The case may be taken as one of pure torsion.

(U. E. I.)

Solution

From Fig. 218, the mean radius of the spring R $= 2d$. Then torque WR $= 300 \times 2d = 600d$. lb. in.

From $\qquad \delta = R\theta, \ \theta = \delta/R = \dfrac{0\cdot5}{2d} = \dfrac{0\cdot25}{d}$ radians.

The length of spring $l = 2\pi RN = 2\pi 2d10 = 40\pi d$ in.

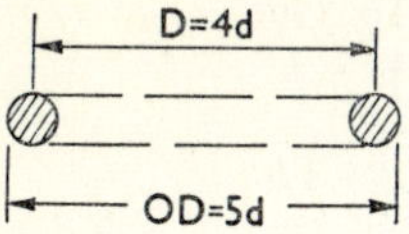

FIG. 218

Applying $\qquad \dfrac{T}{J} = \dfrac{G\theta}{l}$ and substituting $J = \dfrac{\pi d^4}{32}$,

we get $\qquad \dfrac{600d}{\dfrac{\pi d^4}{32}} = \dfrac{12 \times 10^6}{40\pi d} \cdot \dfrac{0\cdot25}{d}$,

rearranging, $\qquad d = \dfrac{32 \times 600 \times 40}{12 \times 10^6 \times 0\cdot25}$

$$d = 0\cdot256 \text{ in.}$$

Example 85

A vertical close-coiled helical spring is suspended at its upper end and supports a weight of 36 lb. from the lower end. The spring is 2·25 inches external diameter and has 15 turns, and the diameter of the wire is 0·25 inch. If the spring stiffness is 48 lb. per inch calculate :
(a) the modulus of rigidity of the steel wire ;
(b) the load the spring would have to carry if the maximum shear stress in the wire were 40,000 lb./in.^{2}

Solution

Mean coil diameter D $= 2\cdot25 - 0\cdot25 = 2\cdot0$ in.
Torque $\qquad$ T $=$ WR $= 36 \times 1 = 36$ lb. in.
Length of spring $\qquad = 2\pi RN = 2\pi \cdot 1 \cdot 15 = 30\pi$ in.

Spring stiffness $\qquad$ S $= \dfrac{W}{\delta}$.

Then deflection $\qquad \delta = \dfrac{W}{S} = \dfrac{36}{48} = 0\cdot75$ in.

Also deflection $\qquad \delta = R\theta.$

Then angle of twist $\theta \;=\dfrac{\delta}{R} = \dfrac{0\cdot75}{1} = 0\cdot75$ radian.

(a) To determine the modulus of rigidity we have by rearrangement of

$$\frac{T}{J} = \frac{G\theta}{l}$$

$$G = \frac{Tl}{J\theta}$$

Substituting the values calculated above

$$G = \frac{36\,.\,30\pi}{\dfrac{\pi}{32}\,.\,\left(\dfrac{1}{4}\right)^4\,.\,0\cdot75} \qquad \therefore \; \mathbf{G = 11\cdot8 \times 10^6\ lb./in.^2}$$

(b) The torque associated with a maximum stress of 40,000 lb./in.^{2} is given by

$$T = \frac{fJ}{r}$$

$$= \frac{40000}{\tfrac{1}{8}}\,.\,\frac{\pi}{32}\,.\,\left(\frac{1}{4}\right)^4$$

$$\therefore \; T = 122\cdot5 \text{ lb. in.}$$

But torque $T = WR$, and since the mean radius $R = 1$ inch, the load to give this value of stress is $\mathbf{W = 122\cdot5\ lb.}$

EXERCISES IX

1. Establish the formulae used in the calculation of torsional effects in circular shafts, stating the assumptions made in their derivation.

A torsion specimen of brass had a diameter of 0·5 inch and a gauge length of 18 inches. Under test the limit of proportionality was reached at a twisting moment of 80 lb. in., when the angle of twist over the gauge length was 2·35 degrees. Calculate the modulus of rigidity for the material. I.Mech.E.

2. Explain the meaning of the term 'Modulus of Rigidity'. A uniform bar of mild steel, 1 inch diameter, was tested within the elastic limit in a torsion machine and gave the following results :

Torque lb. in.	0	200	400	600	800	1000	1200
Angle of twist degrees	0	0·070	0·159	0·229	0·314	0·385	0·467

Plot these results and determine the value of the modulus of rigidity for the material, if the gauge length was 8 inches. I.Prod.E.

3. When under load, a 3 inch diameter solid steel shaft twists through an angle of 8 degrees over a length of 20 feet. If the modulus of rigidity for steel is 12×10^6 lb./in.^{2}, determine the applied twisting moment and the maximum shearing stress in the shaft.

If the shaft rotates at 150 rev./min., determine the horse power being transmitted. U.E.I.

4. A torsion specimen of steel, 0·75 inch diameter, showed that the elastic limit shear stress was reached when the torque applied was 1750 lb. in. At this condition the angle of twist on a length of 8 inches was 2·3°. Determine the modulus of rigidity and the elastic limit shear stress for this material.

Also calculate the minimum diameter of a solid shaft of this material which is required to transmit 80 h.p. at 480 rev./min., if the maximum shear stress in the shaft is not to exceed one-third of the elastic limit shear stress. U.L.C.I.

5. A solid steel shaft, 4 inches outside diameter, transmits a constant horse power 'H'. If the angle of twist for the shaft is limited to 1 degree in 10 feet, determine the maximum shear stress in the shaft and the value of H, if the shaft makes 100 r.p.m.
($G = 5000$ tons/in.^{2})

6. The working conditions to be satisfied by a shaft transmitting power are (a) that the shaft must not twist more than 1 degree on a length of 16 diameters, and (b) the shear stress must not exceed 4 tons/in.^{2} If $G = 12 \times 10^6$ lb./in.^{2}, what is the actual working stress and the diameter of shaft to transmit 1250 h.p. at 240 rev./min. ?

7. Deduce, from first principles, the expression for the polar second moment of area of an annular section whose outside and inside diameters are D and d respectively.

Determine the horse power which can be transmitted at 200 rev./min. by a hollow steel shaft, 12 inches outside diameter and 6 inches internal diameter, if the allowable shear stress is 4 tons/in.^{2} N.C.T.E.C.

8. A machine spindle, 3 inches outside diameter and 2·5 inches inside diameter, running at 400 rev./min. transmits 3 horse power.

(a) Show that the Polar Moment of Inertia for the spindle J
=4·125 in.⁴

(b) Calculate :

       (i) Torque transmitted in lb. in.

       (ii) Maximum shear stress.

       (iii) Angle of twist on a 3-foot length of spindle.

Take Modulus of Rigidity $G = 12 \times 10^6$ lb./in.²   I.Prod.E.

9. Prove the formula $q/r = T/J = G\theta/L$ for a circular shaft and state clearly the meaning of each of the symbols.

The propeller shaft of a car is in the form of a hollow tube, 2 inches diameter and $\frac{1}{8}$ inch thick. Determine the maximum shear stress in the tube when the shaft transmits 70 horse power at a speed of 4000 rev./min.   U.E.I.

10. A hollow shaft, 6 inches external diameter and 4 inches internal diameter, rotates at a speed of 300 rev./min. If the maximum shear stress in the material is limited to 6 tons/in.², calculate the greatest horse power which may be transmitted.

Calculate also the angle of twist of the shaft over a length of 10 feet. Assume $G = 12 \times 10^6$ lb./in.²   U.E.I.

11. A shaft is subjected to a torque T. Deduce the relationship between T, J, q and r, where J is the polar moment of inertia of the circular cross-section of the shaft and q the shear stress produced at a radius r.

Determine the external diameter of a hollow shaft to transmit 2000 h.p. at 210 rev./min., the maximum allowable shear stress being 3 tons/in.² The internal diameter of the shaft is 0·6 of the external.   U.L.C.I.

12. A hollow steel shaft, 6 inches external and 4 inches internal diameter, is to drive a 5000 kilowatt generator at 2500 rev./min. Find :

(a) using a factor of safety of 8, the least shear strength required of the steel ;

(b) the angle of twist of the shaft in a length of 20 diameters, assuming the modulus of rigidity of the steel to be 12×10^6 lb./in.²

(1 H.P. = 746 watts.)

13. Determine the external diameter of a hollow shaft required to transmit 10,000 h.p. at 50 r.p.m., if the maximum allowable shear stress is 3 tons/in.² and the internal diameter is 0·5 of the external diameter. What will be the angular deflection of the shaft in degrees in a length of 10 feet while the horse power is being transmitted ?

(Modulus of rigidity = 12,000,000 lb./in.²)   N.C.T.E.C.

14. A hollow steel shaft, 10 inches outside diameter and 8 inches inside diameter, transmits 5000 h.p. at 300 rev./min. Calculate the shear stress in the outer fibres of the shaft.

Show that for the same skin stress a solid shaft will give a twist $1 \cdot 19$ times greater than the hollow shaft for the same length and torque.

15. A hollow propeller shaft transmits 2500 horse power at a speed of 120 rev./min. The internal diameter of the shaft is two-thirds of the external diameter and the maximum allowable shear stress in the material of the shaft is limited to 4 tons/in.^{2}

Determine :

(a) the external diameter of the shaft, and

(b) the angle of twist over a length of 16 feet, if the shear modulus is 12×10^6 lb./in.^{2} U.E.I.

16. An automobile transmission shaft is required to transfer 30 h.p. at 250 r.p.m. The outside diameter must not exceed 2 inches, and the maximum shear stress is not to be greater than $5 \cdot 3$ tons/in.^{2} Compare the weights of hollow and solid shafts which would just meet these requirements.

17. A hollow steel shaft, 12 inches outside diameter and 8 inches inside diameter, transmits 6000 h.p. at 400 rev./min. Calculate the shear stress in the inner and outer fibres of the shaft.

Show that for the same skin stress a solid shaft will give a twist $1 \cdot 076$ times greater than the hollow shaft for the same length and torque.

18. A steel shaft is required to transmit 80 h.p. at 60 rev./min., and the maximum twisting moment is 25 per cent greater than the mean. Find the diameter of a solid shaft for a maximum allowable shear stress of 7500 lb./in.^{2} What diameter of hollow shaft would be required for this purpose, assuming that the outside diameter is $1\frac{1}{2}$ times the inside diameter ?

19. A hollow steel shaft is required to transmit 25 h.p. at 105 rev./min. If the maximum shear stress is limited to 10,500 lb./in.^{2}, internal diameter $= 0 \cdot 8 \times$ external diameter, and the maximum torque $= 1 \cdot 25$ mean torque, determine suitable values for the external and internal diameters of this shaft.

If the modulus of rigidity for steel $= 12 \times 10^6$ lb./in.^{2}, find the maximum angular twist on a 5 feet length of this shaft.

20. A hollow shaft having internal and external diameters of 2 and $3 \cdot 5$ inches respectively transmits 2000 h.p. when revolving at 2000 rev./min. The shaft is connected to another shaft by a simple flange coupling having eight bolts on a pitch circle of 7 inches diameter. Determine the minimum permissible diameter these bolts may possess if the shear stress in the bolts is not to exceed 7500 lb./in.^{2}

Calculate the maximum shear stress in the shaft under normal working conditions.
U.E.I.

21. Two lengths of hollow shafting, outside diameter 4 inches and internal diameter 2 inches, are connected by a flanged coupling having 6 bolts on an 8-inch P.C.D. If the shaft transmits 180 h.p. at 250 rev./min., determine the size of the bolts required if their average stress is not to be greater than the maximum stress in the shaft. Give bolt diameters to nearest $\frac{1}{8}$ inch. Also determine the stress on the inner surface of the shaft.

22. A hollow shaft, with an internal diameter one-half of the external, is to be designed to transmit 3500 h.p. at 150 rev./min., on the condition that the angle of twist is not to exceed 1 degree per 10 feet of length. Determine the external diameter necessary to satisfy the requirements.

(Modulus of rigidity $=12 \times 10^6$ lb./in.^{2})

Two lengths of this shaft are to be connected by a flanged coupling, which is to have 8 bolts on a pitch circle diameter 1·7 times the shaft external diameter. If the shear stress in the bolts is not to be greater than 10,000 lb./in.^{2}, determine a suitable bolt size.
I.Mech.E.

23. A shaft is 2 inches diameter for part of its length and 1 inch diameter for the remainder. If the shear stress in the material must not exceed 3 tons/in.^{2}, calculate the maximum horse power which can be transmitted by the shaft at 500 rev./min. Calculate the angle of twist per foot length in the 2 inches diameter portion when transmitting this horse power.

(G $=6500$ tons/in.^{2})

24. A solid steel shaft of 3 inches diameter is subject to a maximum torque that causes an angular twist of 5 degrees in a length of 8 feet. It is proposed to substitute a hollow shaft of greater torsional stiffness to reduce the twist to 2 degrees in the same length and with the same torque. The ratio of the internal diameter to the external is to be 0·6. Determine suitable dimensions and calculate the maximum shear stresses in the two cases.

(Modulus of rigidity $=12 \times 10^6$ lb./in.^{2})   I.Mech.E.

25. A shaft, ABC, rotates at 600 r.p.m. and is driven through a coupling at the end A. At B a pulley takes off two-thirds of the power, the remainder being absorbed at C. The part AB is 4 feet long and 4 inches diameter; BC is 5 feet long and 3 inches diameter. The maximum shear stress set up in BC is 6000 lb./in.^{2} Determine the maximum stress in AB and the power transmitted by it; and calculate the total angle of twist in the length AC.

(Rigidity modulus $=12 \times 10^6$ lb./in.^{2})   I.Mech.E.

26. A shaft, ABC, to transmit power at a speed of 300 r.p.m. is to be in two parts, AB and BC, coupled together at B. Power is supplied at B, and the shaft BA of solid section and 5 feet long is to transmit 300 h.p., while BC of hollow section, with an internal bore diameter of 3 inches, is to be 7 feet long and to transmit 350 h.p. Determine the shaft diameters necessary, if the shear stress in BA is not to exceed 5000 lb./in.² and the angle of twist in BC is to be equal to that in BA.

(Modulus of rigidity $= 12 \times 10^6$ lb./in.²) I.Mech.E.

27. A close-coiled helical spring has 20 coils of mean diameter $2\frac{1}{2}$ inches. The diameter of the wire is $\frac{1}{4}$ inch.

If $G = 5200$ tons/in.², determine the extension of the spring due to a load of 40 lb.

28. A close-coiled helical spring has a mean diameter of 4 inches and 20 working coils made from 0·5 inch diameter steel of Modulus of rigidity $= 12 \times 10^6$ lb./in.² Calculate : (a) the stiffness of the spring ; (b) the load on the spring, if the shear stress is 80,000 lb./in.²

29. A close-coiled helical spring made of circular wire is required to absorb 20 ft.lb. of energy for a deflection of 5 inches and a stress not exceeding 12,000 lb./in.² Determine a suitable diameter and length of wire, given that the mean coil diameter is 6 inches.

($G = 11·6 \times 10^6$ lb./in.²)

30. A close-coiled helical spring is required to carry a load of 10 lb. with a deflection of 1 inch, the maximum permissible stress for this deflection being 15,000 lb./in.², and G for the material 12×10^6 lb./in.² If the mean coil diameter may be taken as 10 times the wire diameter, determine the required number of coils and the wire diameter.

Chapter X

HYDROSTATICS

71. Distinction between Fluids and Solids.—The distinction between water and, say, a piece of iron as examples of a fluid and a solid is very self-evident, although in the case of many of the new materials which can exist in different forms at different temperatures and pressures such distinction may not be so apparent.

In such cases it should be remembered that when a solid is subjected to a shear force—as in the case of a loaded beam—distortion occurs and the internal resistance offered by the beam balances the external shear force. When the external force is removed the body returns to its original shape. That quite the opposite occurs for a fluid is well known. A shear force, however small, when applied to a fluid surface, causes the fluid to flow, and it does *not* regain its original shape when the force is removed. Hence the distinction between solid and liquid—a solid can support shear forces, whereas a fluid is incapable of withstanding tangential forces. It is for this latter reason that the force exerted on any surface in a fluid at rest is always perpendicular to that surface.

72. Density, Specific Weight and Specific Gravity.—A clear distinction must be made between density and specific weight if confusion is to be avoided in later studies.

Density is mass per unit volume and is given the symbol ρ (rho):

$$\text{Density } \rho = \frac{\text{Mass}}{\text{Unit Volume}}.$$

Weight, as we already know, is the gravitational force exerted on a body and therefore:

Specific Weight refers to the gravitational force per unit volume and has the symbol w. Consistent with the engineers' units already employed in previous chapters the unit of mass is the Slug—i.e. 32·2 pounds—and the unit of force is the pound weight.

We then have : $\rho = \dfrac{\text{Slugs}}{\text{ft.}^3}$ and $w = \dfrac{\text{lb.}}{\text{ft.}^3}$.

The relationship between w and ρ is easily understood since specific weight is the gravitational force which acts on a mass of unit volume to produce an acceleration of g ft./sec./sec.

Hence $$w\left(\frac{\text{lb.}}{\text{ft.}^3}\right) = \rho\left(\frac{\text{Slugs}}{\text{ft.}^3}\right)g$$

or simply $w = \rho g$ numerically.

The specific weight of fresh water will be taken as 62·4 lb./ft.³, and, therefore, the density of water :

$$\rho = \frac{62\cdot4}{32\cdot2} = 1\cdot94 \text{ Slugs/ft.}^3$$

If the absolute system of units is employed then it must be remembered that density will have the units pounds per cubic foot and specific weight poundals per cubic foot.

The Engineers' System is to be preferred.

Throughout the following pages it will be observed that w the specific weight is assumed to be the same at all points in a liquid. This is the same as assuming that the liquid is incompressible. It is worth noting, however, that, for gases, the specific weight varies with temperature and pressure to such an extent that the variation must be considered.

Specific Gravity—the ratio of the density or specific weight of a given liquid to the density or specific weight of water (usually at 4°C. and atmospheric pressure) is called the specific gravity of the liquid and is given the symbol s.

73. Intensity of Pressure and Thrust.—If a force or thrust P is applied uniformly over an area A to a liquid in a container the intensity of pressure is defined as :

$$p = \frac{P}{A} \text{ (Compare with definition of stress given in Chapter VI.)}$$

Where P is not applied uniformly across the area this equation then gives an average value.

There can be confusion over the use of the terms pressure and intensity of pressure. Many text-books abbreviate 'intensity of pressure' to pressure and also use the term pressure when total force is clearly intended. To avoid any misunderstanding the term pressure will be used in place of intensity of pressure to mean force per unit area and the term 'thrust' will denote force only.

74. Pressure at a point in a liquid.—It has already been mentioned that the force exerted on any surface in a liquid at rest can only act normal to that surface.

We shall now use this fact to show that the pressure at a point in a liquid is the same in all directions.

Consider a small triangular element of liquid within a body of liquid as shown in Fig. 219.

Let the pressures on the triangular faces be p_1, p_2 and p_3 and the thickness of the element perpendicular to the paper be unity.

Area AB = unity
Area BC = $\sin \theta$
Area AC = $\cos \theta$

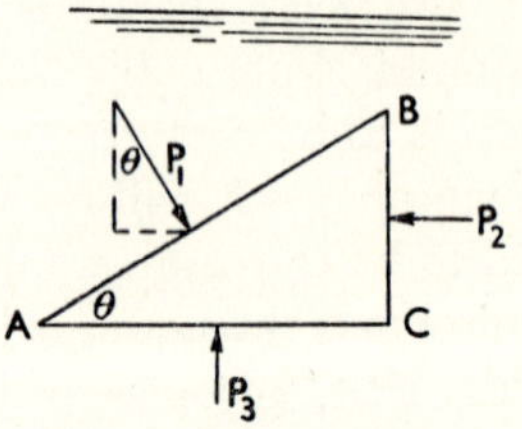

Thrust $P_1 = p_1 \times$ unity
$P_2 = p_2 \sin \theta$
$P_3 = p_3 \cos \theta$

Fig. 219

Then for equilibrium in a liquid at rest the net horizontal forces on the liquid element are zero and the net vertical forces on the element are also zero

or	$p_1 \times$ unity $\times \sin \theta = p_2 \sin \theta.$	
Hence	p_1	$= p_2$
and	$p_1 \times$ unity $\times \cos \theta = p_3 \cos \theta.$	
Hence	p_1	$= p_3.$
Therefore	$p_1 = p_2$	$= p_3.$

Thus the pressure at a point in a liquid is the same in all directions since the above result is independent of the inclination of P_1.

75. Relationship between Pressure and Head.—Imagine an elemental cylinder of liquid length L within the main body of a liquid at rest as shown in Fig. 220.

Let it be assumed for the present that the free surface is a surface of zero pressure.

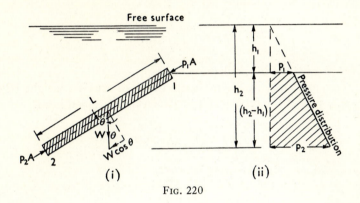

FIG. 220

Then for equilibrium of the cylinder we know that the resultant force in *any* direction must be zero. We shall consider the uniform cross-sectional area A to be small enough for the pressures p_1 and p_2 on the ends to be uniformly distributed.

For equilibrium along the axis of the cylinder there are only three forces to be considered :

(i) Thrust p_1A acting at (1).
(ii) Thrust p_2A acting at (2).
(iii) The component of the total weight W of the cylinder along the axis equal to $wAL \cos \theta$, where AL is the cylinder volume and w the specific weight of the liquid.

Then $\qquad p_2A - p_1A - wAL \cos \theta = 0,$

or $\qquad\qquad\qquad p_2 - p_1 = wL \cos \theta,$

but $\qquad\qquad\qquad L \cos \theta = h_2 - h_1,$

and, therefore, $\qquad\qquad p_2 - p_1 = w(h_2 - h_1).$ . . (1)

At a point in the free surface $h_1 = 0$, and as we have our assumed surface of zero pressure $p_1 = 0$, therefore, substituting in equation (1) $p_2 = wh_2$, and this expression gives the pressure corresponding to any depth below the free surface.

This clearly indicates that the pressure within a liquid varies linearly with the depth or 'head' of liquid as in Fig. 220 (ii).

In general, $p = wh.$.   .   .   (2)

Units.—It is usual to measure depth or head h in feet and specific weight in pounds per cubic foot.

Hence $p = w \left(\dfrac{\text{lb.}}{\text{ft.}^3} \right) h(\text{ft.}),$

gives pressure in lb. per square foot.

For pressures in lb./in.² with water of specific weight 62·4 lb./ft.³, we get the relationship between pressure and head.

$$p \left(\frac{\text{lb.}}{\text{in.}^2} \right) = \frac{62 \cdot 4}{144} \, h \, (\text{ft.})$$

or $p \, \text{lb./in.}^2 \equiv \dfrac{h \, \text{ft.}}{2 \cdot 308}$

which shows that each 2·308 feet increase in depth of water corresponds to a change of 1 lb./in.² pressure.

We thus have in this relationship between p and h an alternative means of indicating the pressure at any point in a liquid.

From equation (2) we have $h = \dfrac{p}{w}$, and in this connection $\dfrac{p}{w}$ is often called the pressure head.

76. Measurement of Pressure.—Most students at this stage will have used simple measuring instruments—if not in industry at least in the college laboratory—and, therefore, will be acquainted with the terms absolute, gauge and vacuum as applied to pressure measurement.

We are normally interested in pressures relative to the pressure of the atmosphere taken as the datum, and instruments of the Bourdon gauge type give pressures above atmospheric, i.e. 'gauge' pressure.

Pressures measured from the true zero of pressure are called 'absolute' pressures.

Hence Absolute Pressure = Gauge Pressure + Atmospheric Pressure.

If atmospheric pressure is taken as the datum from which

measurements are to be made, then it is readily seen from Fig. 221 that 'pressures' can exist above and below atmospheric pressure.

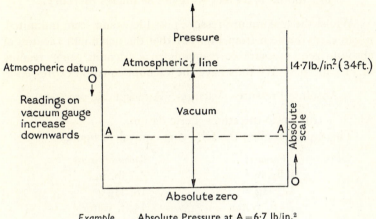

Example.   Absolute Pressure at A = 6·7 lb/in.²
Corresponding reading
on Vacuum Gauge   = 8 lb/in.²

FIG. 221

For this reason it will be understood why these are often referred to as positive and negative pressures, although there is no such thing as a negative pressure—all pressures in fact being truly positive above absolute zero.

The positive and negative notation for pressures should be avoided. Instead it is better to use the term 'vacuum' to indicate that pressures below atmospheric are under consideration and to reserve the term pressure for conditions above atmospheric.

Using 14·7 lb./in.² as standard atmospheric pressure in the equation $p = wh$ for water, we have

$$14·7 \times 144 = 62·4 \, h,$$
$$\therefore h = 34 \text{ ft.}$$

Hence standard atmospheric pressure is equivalent to 34 feet of water. It is usual to say that the atmosphere will support a water column of 34 feet head.

Similarly, since mercury has an approximate specific gravity of 13·6, then 34 feet of water is equivalent to $\dfrac{34 \times 12}{13·6} = 30$ inches of mercury. We, therefore, have the following alternative methods of expressing atmospheric pressure :

14·7 lb./in.² or 34 feet water or 30 inches mercury.

Whilst the maximum pressure possible would seem unlimited, it can easily be seen from Fig. 221 that the maximum vacuum or absolute zero pressure corresponds to a vacuum gauge reading of 14·7 lb./in.². We thus have the vacuum equation expressed as

Absolute Pressure + Vacuum = Atmospheric pressure,

which is true up to the atmospheric datum.

This equation gives the following equivalent figures :

Vacuum	*Absolute Pressure*
8 lb./in.²	6·7 lb./in.²
12″ Hg	18″ Hg
20 ft. water	14 ft. water

Not all pressure measurements are made with Bourdon gauges, and often use is made of manometer or U-tube instruments. The U-tube is illustrated in Chapter XI in connection with the flow through a Venturi meter and consists simply of a U-shaped tube filled with a measuring liquid—usually mercury—which is moved by an amount proportional to the pressure being measured.

Example 86

A uniform upright tube, 0·5 inches diameter, contains 6 inches of water above a 2-inch length of mercury. What is the pressure in lb./in.² at the base of the tube ? Specific gravity of mercury 13·6.

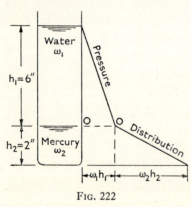

Fig. 222

Solution

The pressure at the common surface OO is due to the 6-inch column of water and this pressure will be transmitted through the mercury as shown by the pressure distribution in Fig. 222.

Total pressure at the base $= w_1 h_1 + w_2 h_2$ lb./ft.²

$$= \left(62 \cdot 4 \times \frac{6}{12} + 62 \cdot 4 \times 13 \cdot 6 \times \frac{2}{12}\right) \text{ lb./ft.}^2$$

$$= 172 \cdot 5 \text{ lb./ft.}^2$$

Hence pressure at the base $= \textbf{1·196 lb./in.}^2$

It is seen that the diameter of the tube does not appear in the calculation. This is because 'pressure' is concerned with 'head' only and hence the diameter is irrelevant to this question.

Example 87

When one leg of an open mercury U-tube is connected to the centre of a pipe containing water under pressure, the mercury column is deflected 6 inches. If the centre of the pipe is 1 foot below the level of the lower mercury meniscus, what is the pressure head in the pipe ? If the lower mercury meniscus is now made to coincide with the centre of the pipe, how is the manometer reading affected ?

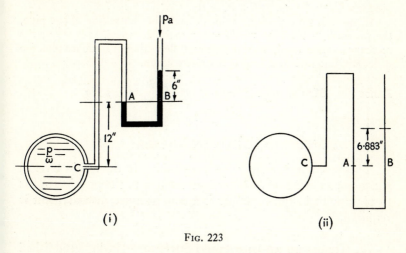

(i)   (ii)

Fig. 223

Solution—Fig. 223

Let pressure head h in pipe $= \dfrac{p}{w}$ ft. of water.

Now for Equilibrium (since mercury column is continuous),

Pressure at A = Pressure at B.

But pressure at B is that due to a Hg column 6 inches long plus atmospheric pressure.

$$\therefore \text{ Pressure at A} = \frac{6}{12}13\cdot6 + 34 \text{ feet water}$$

$$= 6\cdot8 + 34 = \textbf{40·8 feet water.}$$

Now pressure at centre of pipe supports pressure at A together with a column of water of length CA = 1 foot.

Hence pressure head in pipe

$$\frac{p}{w} = \text{Pressure at A} + 1 \text{ foot}$$

$$= 40\cdot8 + 1.$$

$$\therefore \frac{p}{w} = \textbf{41·8 ft. water.}$$

Since we have included atmospheric pressure in the calculation this is the absolute pressure in the pipe.

The pressure recorded by a Bourdon gauge placed central with the pipe would, therefore, be 7·8 ft. water.

If, now, A were made to coincide with the centre line of the pipe, the pressure in the pipe would still remain the same. However, the mercury column would now have a total deflection of 6 inches plus the deflection due to 1 foot length of water column.

Hence $$\text{Manometer reading} = 6 \text{ inches} + \frac{30}{34} \text{ inches Hg}$$

$$= 6 + 0\cdot883$$

Manometer reading = **6·883 in. Hg.**

This example indicates that the position of the liquid level in a manometer must be considered in relation to the position of the tapping point in the container if a true pressure measurement is required.

77. Thrust on an Immersed Surface.—Having established the variation of pressure with depth and also that the thrust exerted on any surface in a liquid is always perpendicular to that surface, we can use these principles to determine the resultant thrust exerted on an immersed surface.

Consider a uniform plate immersed in a liquid as shown in Fig. 224.

O is the intersection of the plane of the plate continued with the free surface and makes an angle θ with it.

For an elemental strip of the surface distance y from O and thickness dy we have :

Area of Element $= b\,dy$
Normal pressure on element $p = wh$
Then thrust on element $= wh \,.\, b \,.\, dy$ (pressure × area)
But $h = y \sin \theta$
$\therefore$ thrust on element $= w \sin \theta \, b\,dy \,.\, y \,.$.   (1)

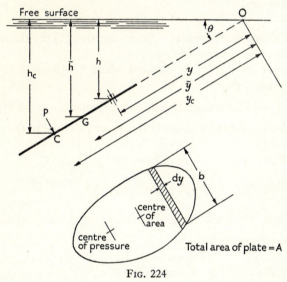

Fig. 224

Now the total thrust exerted on the whole plate will be the sum of all such elemental thrusts.

Hence Total Thrust on Plate $= \int w \sin \theta \, b\,dy \,.\, y.$

But $b\,dy \,.\, y =$ first moment of area of the
                                                         element.

$\therefore \int b\,dy \,.\, y$ is clearly the first moment of area of the whole plate about O, i.e. $A\bar{y}$ where $\bar{y}$ is the distance of the centre of area from O. (See also Chapter VIII.)

Therefore $P = w \sin \theta \, A\bar{y}.$.   .   .   (2)

Replacing $\bar{y} \sin \theta$ by $\bar{h}$—the depth of the centre of area below the free surface—we get the final expression :

Resultant thrust $\qquad P = Aw\bar{h}. \qquad . \qquad . \qquad . \qquad (3)$

We have, therefore, shown that the resultant thrust on an immersed surface is given by the product of the pressure at the centre of area multiplied by the total area of surface. The resultant thrust is thus independent of the inclination of the immersed surface provided its centre of area remains at the same depth. It remains for us to establish where the resultant thrust may be assumed to act.

78. Centre of Pressure.—From equation (1) the thrust on the elemental area $= w \sin \theta \, bdy \, . \, y$.

Moment of thrust on element *about* $O = w \sin \theta \, bdy \, . \, y \, . \, y$
$= w \sin \theta \, bdy \, . \, y^2$.

The Total Moment of Thrust for all elements $= w \sin \theta \int bdy \, . \, y^2$.

But for equilibrium about O this must be equal to the moment of P about O.

Let P act at the 'centre of pressure' y_c from O.

Then $\qquad\qquad Py_c = w \sin \theta \int bdy \, . \, y^2$.

Since $\int bdy \, . \, y^2$ is the Second Moment of Area I_0 of the plate about O, we get

$$Py_c = w \sin \theta \, I_0.$$

But from (2) $\qquad P = w \sin \theta \, A\bar{y}$,

$\therefore \; w \sin \theta \, A\bar{y} \, . \, y_c = w \sin \theta \, I_0$

or $\qquad\qquad\qquad y_c = \dfrac{I_0}{A\bar{y}} = \dfrac{\text{Second Moment of Area}}{\text{First Moment of Area}}$.

It is essential to remember that in this expression both the first and second moments of area are calculated about O as axis.

Alternatively, by using the parallel axis theorem :

$$I_0 = I_G + A\bar{y}^2,$$

where I_G is the second moment of area about the centroid.

Then $\qquad\qquad y_c = \dfrac{I_G + A\bar{y}^2}{A\bar{y}} = \dfrac{I_G}{A\bar{y}} + \bar{y}.$

This clearly shows that the 'centre of pressure' is below the centroid of the section by an amount $\dfrac{I_G}{A\bar{y}}$.

A further simplification using $I_G = AK_G^2$, where K_G is the radius of gyration, yields :

$$y_c = \frac{K_G^2}{\bar{y}} + \bar{y}.$$

Again if we use $h = y \sin \theta$, we obtain :

$$y_c = \frac{K_G^2}{\bar{h}} \sin \theta + \frac{\bar{h}}{\sin \theta}.$$

If the plate is vertical, then $\sin \theta = 1$ and $y_c = h_c$,

$$\therefore \; h_c = \frac{K_G^2}{\bar{h}} + \bar{h},$$

where $\bar{h}$ is the vertical depth below the free surface to the centre of area.

Two important cases—
(i) Rectangular plate with edge in free surface.

Consider the equation $\qquad y_c = \dfrac{I_0}{A\bar{y}}.$

For a rectangular vertical surface with its top edge in the water surface $\theta = 90°$ and $y_c = h_c$ (see Fig. 224).

Also I_0 corresponds to I about the base of a rectangle. Hence for the plate in Fig. 225 (i)

$$I_0 = \frac{bd^3}{3}, \; A = bd, \; \bar{y} = \bar{h} = \frac{d}{2}.$$

Then $\qquad\qquad y_c = h_c = \dfrac{bd^3}{3} \bigg/ bd \cdot \dfrac{d}{2}$

$$y_c = h_c = \frac{2}{3} d,$$

or the depth of the centre of pressure is $2/3$ the length of the vertical below the free surface.

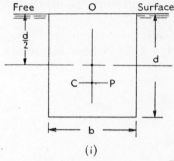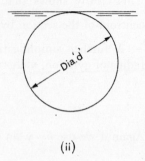

FIG. 225

(ii) Circular plate with edge in free surface. Fig. 225 (ii).
Using the parallel axis equation

$$y_c = \frac{I_G + A\bar{y}^2}{A\bar{y}} = \frac{I_G}{A\bar{y}} + \bar{y}$$

we have

$$I_G = \frac{\pi d^4}{64}, \ A = \frac{\pi d^2}{4}, \ \bar{y} = \frac{d}{2}.$$

Hence

$$= \frac{\dfrac{\pi d^4}{64}}{\dfrac{\pi d^2}{4} \times \dfrac{d}{2}} + \frac{d}{2} = \frac{d}{8} + \frac{d}{2},$$

or y_c the depth of the centre of pressure is $\dfrac{5d}{8}$ below the free
surface.

Example 88

A sluice gate is 16 feet wide and has water to a depth of 24 feet on
one side and 12 feet on the other. Find the resultant horizontal thrust
on the gate and the position of its line of action from the bottom.

Solution—Fig. 226

Horizontal thrust on deep side

$$\begin{aligned}
P_1 &= wA_1\bar{h}_1 \\
&= 62 \cdot 4 \times (16 \times 24) \times 12 \\
&= 288,000 \text{ lb.}
\end{aligned}$$

Horizontal thrust on shallow side

$$P_2 = wA_2\bar{h}_2$$
$$= 62{\cdot}4 \times (16 \times 12) \times 6$$
$$= 72{,}000 \text{ lb.}$$

Therefore resultant horizontal thrust on the gate R

$$= 288{,}000 - 72{,}000,$$
$$= \textbf{216,000 lb.}$$

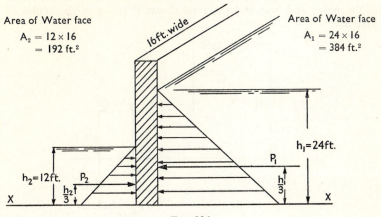

Area of Water face
$A_2 = 12 \times 16$
$= 192 \text{ ft.}^2$

Area of Water face
$A_1 = 24 \times 16$
$= 384 \text{ ft.}^2$

16ft. wide

$h_1 = 24\text{ft.}$

$h_2 = 12\text{ft.}$ P_2

P_1

$\dfrac{h_2}{3}$

$\dfrac{h_1}{3}$

X   X

FIG. 226

Thrust P_1 acts at $\dfrac{h_1}{3} = 8$ ft. from XX

Thrust P_2 acts at $\dfrac{h_2}{3} = 4$ ft. from XX

Moment of P_1 about XX $= 288{,}000 \times 8 = 2{,}304{,}000$ lb. ft.
Moment of P_2 about XX $= 72{,}000 \times 4 = 288{,}000$ lb. ft.

Let resultant thrust R act at distance y from XX.

Then $216{,}000 \times y = 2{,}304{,}000 - 288{,}000$

$$= \dfrac{2{,}016{,}000}{216{,}000}.$$

Resultant thrust acts at $y = \textbf{9·34 feet from XX.}$

Example 89

A flap valve, 2 feet diameter, closes the end of a horizontal pipe against internal water pressure, the head of water over the centre line of the pipe being 5 feet. The valve is held by a hinge, whose pin is 1 foot 6 inches above the centre of the flap, and by a bolt 1 foot 6 inches below the centre of the flap. Determine the forces on the hinge and the bolt.

Solution

It is clear that the flap valve will be in equilibrium under the forces at the hinge and bolt and the total thrust P acting at the centre of pressure.

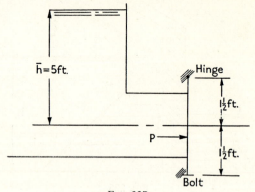

Fig. 227

$$\text{Total thrust } P = wAh$$

$$= 62.4 \times \frac{\pi}{4} . 2^2 . 5,$$

$$= \mathbf{980 \ lb.}$$

This thrust acts at the centre of pressure, which is at a distance below the free surface given by $\bar{h} + \dfrac{I_G}{A\bar{h}}$, or, alternatively, $\dfrac{I_G}{A\bar{h}}$ below the centre line of the pipe.

For a circular disc $\qquad\qquad I_G = \dfrac{\pi d^4}{64}.$

Hence centre of pressure is $\dfrac{\pi d^4}{64} \bigg/ \dfrac{\pi d^2}{4} . 5$ below centre line of pipe,

i.e. $\qquad \dfrac{d^2}{16 \times 5} = \dfrac{4}{80} = 0.05$ ft.

To find the force on the bolt we take moments about the hinge.

Therefore $F_B \times 3 \text{ ft.} = P \times (1\frac{1}{2} + 0.05)$

$$F_B = \frac{980 \times 1.55}{3} = \textbf{506 lb.}$$

Since total thrust on valve $= 980$ lb. Force on hinge $F_H = 980 - 506$

$$F_H = \textbf{474 lb.}$$

Example 90

A storage tank is to have sides sloping at 30° to the horizontal and retain water to a depth of 5 feet. The sides are to be supported at 6-foot intervals by timber struts perpendicular to them. These struts are to meet the tank at a point 6 feet from the ground along the sloping side. If the timber has a UTS of 10,000 lb./in.², determine a suitable diameter for the struts, using a factor of safety of 5.

Solution—Fig. 228

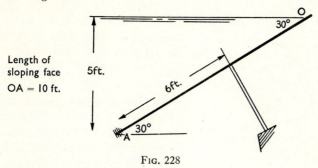

Length of
sloping face   5ft.
OA = 10 ft.

30°

6ft.

30°

O

A

FIG. 228

Area of sloping face of tank supported by each strut $= 10 \times 6 = 60$ ft.²

Total thrust on side $P = wA\bar{h}$, where $\bar{h} = 2\frac{1}{2}$ ft. below free surface,
                                $= 62.4 \times 60 \times 2.5$
                                $= 9360$ lb.

Total thrust will act at centre of pressure

$$y_c = \frac{I_G}{A\bar{y}} + \bar{y}$$

from O where y is distance to centroid measured along slope.

A.M.—U

Hence
$$y_c = \frac{6 \times 10^3}{12} \Big/ (6 \times 10 \times 5) \quad +5$$

$$= 1\tfrac{2}{3} + 5 = 6\tfrac{2}{3} \text{ ft. from O.}$$

Now thrust taken by each strut may be found by taking moments about A.

$$\text{Thrust on strut} \times 6 = P \times (10 - 6\tfrac{2}{3})$$

$$\text{Thrust on strut} = \frac{9360 \times 3\tfrac{1}{3}}{6}$$

$$= 5200 \text{ lb.}$$

$$\text{Permissible stress in strut} = \frac{10,000}{5} = 2000 \text{ lb./in.}^2$$

$$\text{Cross sectional area of strut} = \frac{5200}{2000} = 2\cdot6 \text{ in.}^2$$

Therefore
$$d^2 = \frac{4}{\pi} \times 2\cdot6 \text{ and } d = 1\cdot82''.$$

Hence   **$2''$ dia. struts would be suitable.**

79. Archimedes' Principle.

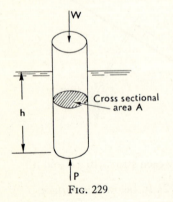

FIG. 229

—When a body of weight W floats in a fluid it is quite easily appreciated that there must be an equal upward fluid reaction equivalent to W acting on the body. That this reaction is due to the weight of the fluid displaced by the body was first stated over 2000 years ago by Archimedes and can readily be verified as follows :

Consider a uniform body of weight W floating upright in a fluid as shown in Fig. 229. Let the depth of immersion be h.

Then at depth h pressure is given by $p = wh$.

Therefore,   thrust $P = pA = whA$.

Now for equilibrium $W = P$.

Hence $W = whA$.

But hA is the volume of displaced fluid and hence whA is the weight of the displaced fluid. It is thus shown that 'when a body is submerged in a fluid it is made buoyant by a force equal to the weight of the fluid displaced'. This is referred to as the principle of Archimedes.

80. Stability.—A body is said to be in 'stable' equilibrium if it returns to its original position after being slightly displaced ; in 'neutral' equilibrium if it remains in the new position after being slightly displaced ; and in 'unstable' equilibrium if it continues to move in the direction of the displacement.

These conditions are illustrated in Fig. 230 for three balls located respectively on concave, flat and convex surfaces.

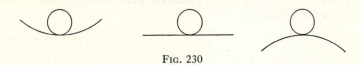

Fig. 230

Fig. 231 (a) shows a body of weight W in stable equilibrium, i.e. the buoyancy thrust P, is in line with and equal to W.

The centre of gravity of the body is shown at G and the centre of buoyancy (the centre of gravity of the displaced fluid) is shown at B.

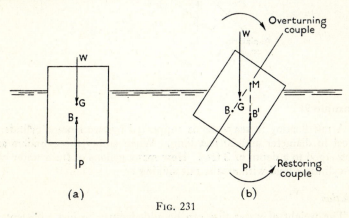

(a)   (b)

Fig. 231

Suppose that for some reason the body is tipped as shown in Fig. 231 (b). Then immediately the buoyancy thrust is displaced to some such position as B¹, although the position of G remains unaltered.

Clearly the equal and opposite thrusts W and P now form a couple which, for this case, tends to restore the equilibrium of the body.

It should be noted that the vertical through the new centre of buoyancy cuts the line BG at a point M. Provided that the list is small, M is called the 'metacentre'. For the case under consideration M lies above G, and stable conditions exist.

If, as in Fig. 232, M lies below G, it will be seen that once the body is tipped the couple introduced aggravates the rolling—i.e. it tends to tip the body still further and the body is said to be unstable.

Hence for stability the metacentre must be *above* the centre of gravity of the body.

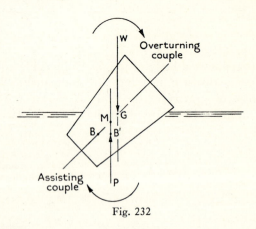

Fig. 232

Example 91

A raft floating in sea water is supported by two storage cylinders, 3 feet in diameter and 20 feet long. When empty the cylinders are submerged to a depth of 2 feet. How many gallons of fresh water can the cylinders hold without the raft sinking ?

Solution

The weight of water that can be added must be just sufficient to submerge the shaded segment of cylinder as shown in Fig. 233.

Now from Archimedes' principle the weight added will equal the thrust exerted by the volume of water displaced by the shaded segment.

Cross-sectional area of segment of one cylinder equals area of sector OABC – Area triangle OAC :

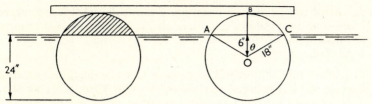

Fɪɢ. 233

$$\theta = \cos^{-1} \frac{6}{18} \qquad \therefore \ \theta \text{ in radians} = 1 \cdot 231.$$

$$\text{Area of Sector} = \frac{r^2}{2} \text{ (Angle of Sector)}$$

$$= \frac{18^2}{2} \cdot 2 \times 1 \cdot 231 = \mathbf{400} \text{ in.}^2$$

Area of triangle OAC $= 6\sqrt{18^2 - 6^2} = \mathbf{102}$ in.^{2}

Hence   Area of Single Segment $= \mathbf{298}$ in.^{2}

Total Cross-sectional area submerged $= \mathbf{596}$ in.^{2}

$$\text{Total Volume displaced} = \frac{596}{144} \times 20$$

$$= \mathbf{82 \cdot 8} \text{ ft.}^3$$

Weight of displaced sea water $= 82 \cdot 8 \times 64$

$$= \mathbf{5300} \text{ lb.}$$

This represents the weight of fresh water that can be added without sinking the raft.

Therefore, **530 gallons** may be carried.

EXERCISES X

1. A tube is filled with water to a depth of 2 feet, and then 18 inches of oil of specific gravity 0·72 is poured in and allowed to come to rest. Determine the pressure (*a*) at the common surface, (*b*) at the base of the tube.

2. One leg of a vertical U-tube containing mercury is connected to the bottom of an open vertical tank, the other leg being open to the atmosphere. When the tank is empty the level of the mercury in the U-tube is 3 feet below a fixed mark on the tank. If water is now poured into the tank up to the level of the fixed mark, determine the resulting difference in the mercury levels in the U-tube.

3. A rectangular tank, 5 feet long, 4 feet high and 2 feet 6 inches broad, is filled with fresh water to a height of 3 feet 6 inches. Determine the force (*a*) on the base of the tank ; (*b*) on one side ; (*c*) on one end of the tank.

4. A lock gate, 20 feet wide, has water on both sides, the depth of water on one side being 15 feet and on the other 9 feet above its base.

Determine the magnitude of the resultant thrust on the gate, and the position at which it acts.

5. A plate of rectangular form is vertically submerged in a liquid with one side in the free surface. Obtain the expressions for the resultant thrust on one side of the plate and the depth to the point at which it acts.

A tank is divided into two compartments by a vertical partition plate. One compartment contains water to a depth of 7 feet ; the other holds oil, of specific gravity 0·94, to a depth of 4 feet. Determine, per foot width of plate, the position and magnitude of the resultant force on the partition.

(Specific weight of water, 62·5 lb./ft.³) I.Mech.E.

6. A circular plate, 2 feet in diameter, is subjected to water pressure on one side only. Working from first principles find the total force in pounds and the position at which it may be assumed to act, measured below the centre of the plate, if the centre of the plate is 3 feet below the surface.

7. An open tank with vertical sides contains water. A hole, 3 feet diameter, with its centre 4 feet below the water surface, is covered by a plate. Determine the resultant force and the point at which this force acts on the plate.

(Density of water is 62.4 lb./ft.³) U.L.C.I.

8. A cylindrical tank, 3 feet in diameter, has its axis horizontal. At the middle of the tank, on top, is a pipe, 2 inches in diameter, which extends vertically upwards. The tank and pipe are filled with oil, of specific gravity 0·97, with the free surface in the 2-inch pipe at a level of 12 feet above the tank top. Determine the total force on one end of the tank.

9. A vertical lock-gate has a rectangular aperture in it 3 feet high × 4 feet wide, closed by a door which overlaps the aperture 6 inches all round. The door is hinged at its top edge. Calculate the force on the hinge and on the fastening at the bottom edge when the water stands 6 feet above the top edge of the aperture, the door and the water being on opposite sides of the lock-gate.

10. Define the terms 'Total Thrust' and 'Centre of Pressure'. A square trap-door of side 3 feet in the vertical side of a water tank has its lower edge horizontal and hinged. It is kept closed by a normal force of 500 lb. applied to the upper edge. Find the greatest depth of water above the lower edge which the tank can contain.

11. A hollow right circular cylinder is submerged in water so that its axis is horizontal and 10 feet beneath the surface. The ends of the cylinder are closed. To what pressure must the air inside the cylinder be raised in order that there shall be no resultant thrust on the circular ends of the cylinder ? If the diameter of the cylinder is 2 feet, calculate the resultant couple on one end. The thickness of the cylinder is negligible.

12. One side of a tank containing water is inclined at 30° to the vertical. A triangular opening in this side is closed by a door in the form of an equilateral triangle of side 3 feet which is hinged horizontally along one edge 6 feet below the free water surface. If the depth of water in the tank is 7 feet, what horizontal force is required at the apex to keep the door closed ?

13. A container is made in the form of a cubical box from sheet metal weighing 22 lb./ft.² The side of the box is 3 feet long, and when the box floats in fresh water the top is horizontal. Determine the least weight that must be placed on top of the box in order to sink it.

Chapter XI

HYDRODYNAMICS

81. Distinction between Fluid and Solid Motion.—When a solid body is in motion, whether it be translation or rotation, the individual particles which make up the body all undergo similar movements. This is apparent, since for a solid body the positions of all the particles are fixed relative to each other. Hence, as we have already shown, if we stipulate the translation of an individual particle of a solid body we automatically define the motion of the body as a whole.

Unfortunately this is by no means true in the case of a fluid for which the individual particles are not fixed relative to each other, and the problem, therefore, arises, how we can best describe the motion of a body of liquid from the separate motions of its individual particles.

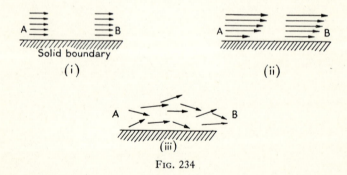

Fig. 234

Let us consider three ways in which a quantity of liquid could flow, say, from A to B in Fig. 234.

At (i) the individual particles are considered to flow in parallel planes all with the same velocity. This type of flow may be taken as applying only to the ideal fluid—i.e. one in which fluid friction is non-existent. Fig. 234 (ii) illustrates what is called laminar flow. Here again the flow is in parallel planes, but conditions are such that the particle velocities increase with the distance from the solid boundary. This type of flow can actually exist, although in the realm of engineering applications it is an exception. It is usually only found in streams of minute thickness, such as in the lubricating films of bearings.

The actual way in which fluid flow is likely to take place is depicted at (iii). This is described as turbulent flow and consists of the random motion of individual particles in the general direction of fluid motion.

82. Rate of Flow.—If the ideal type of flow, as illustrated in Fig. 234 (i), were to occur in a pipe it would be a simple matter to calculate the rate of flow. This can be shown as follows :

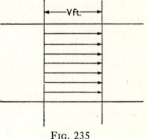

FIG. 235

Let the particles all move with velocity V ft./sec. in a pipe of uniform cross-sectional area A sq. ft.

Then in one second the body of fluid will move V ft. and the volume passing any normal cross-section of the pipe in one second will be AV cubic ft. Fig. 235.

Hence   Volume Flow = AV cubic feet per second,

or   **Quantity = AV Cusec,**

where cusec is an abbreviation for cubic feet per second.

Now, if the specific weight of the fluid flowing is w lb./ft.^{3}, we have

Weight rate of flow W = wAV lb./second.

It would now be natural to inquire how this method of calculating the rate of flow can possibly help us for the case of an actual flow as depicted in Fig. 234 (iii). From the outset it should be clearly appreciated that the accurate prediction of the rate of flow in turbulent motion is impossible. Fortunately, however, results sufficiently accurate for engineering purposes can be obtained by considering the 'average' velocity of all the individual particle velocities along the direction of the pipe axis. In other words, we employ the ideal equation given above, but use an average velocity for all the fluid particles.

Hence   Quantity = Area × Average Velocity
          (ft.³/sec.)   (ft.²)   (ft./sec.)

We, therefore, obtain the average velocity of flow in, say, a pipe or channel by measuring the quantity flowing in a given time and dividing by the normal cross-sectional area of flow.

Therefore, if V now represents the average velocity, we have

$$Q = AV \text{ Cusec}$$
and
$$W = wAV \text{ lb./sec.}$$

83. Equation of Continuity.—Only in the case of a pipe of uniform diameter and full of liquid or a channel of uniform cross-

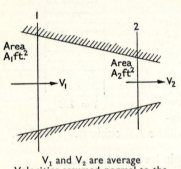

Area A_1ft.²

Area A_2ft.²

V_1 V_2

V_1 and V_2 are average Velocities assumed normal to the section considered.

Fig. 236

sectional area and uniform depth can the average velocity of flow be considered constant at every section. In all other cases it is to be expected that this velocity will vary.

Consider the converging pipe in Fig. 236.

If two normal cross-sections of the pipe are taken, at which the areas are A_1 and A_2 sq. ft., and if the pipe is assumed to run full, then in any given time the quantity of fluid entering section (1) must be equal to the quantity leaving section (2) for a constant rate of flow.

Then, from the equations given above—

          Rate of flow at section (1) = $w_1A_1V_1$ lb./sec.
and   Rate of flow at section (2) = $w_2A_2V_2$ lb./sec.
Hence   from Continuity of flow $w_1A_1V_1 = w_2A_2V_2$. .   . (1)

This equation applies for all fluids—i.e. liquids and gases. For liquids that may be considered as incompressible, w, the specific weight, remains sensibly constant and, therefore, $w_1 = w_2$.

Then   from equation (1) $A_1V_1 = A_2V_2$.   . (2)

This is called the 'equation of continuity' for an incompressible fluid and clearly shows that the average velocity of flow is inversely proportional to the cross-sectional area of the pipe.

84. Pressure and Potential Energy.—We already know that if a particle of weight W is at rest and exists at a height h above a given datum, then relative to that datum it possesses Potential Energy of value Wh ft.lb.

Similarly, for a quantity of fluid weight W lb. at the surface of a tank (Fig. 237), its potential energy above the datum is Wh_1.

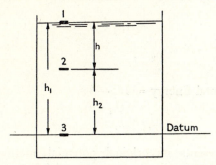

Fig. 237

If this quantity of fluid is moved to a position (3) coinciding with the level of the datum, then clearly it will now possess no potential energy relative to the datum.

However, from the principle of the Conservation of Energy it must still possess energy equivalent to Wh_1 ft.lb. Now we know from the previous chapter that at a depth h_1 we have the relationship $h_1 = \frac{p_1}{w}$, where $\frac{p_1}{w}$ is called the pressure head. Hence the energy at (3) can be written $W\frac{p_1}{w}$, and is called the pressure energy.

I.e. Pressure Energy $= W\frac{p_1}{w}$ ft.lb. or $\frac{p_1}{w}$ ft.lb. per lb. of fluid.

N.B.—$\frac{p_1}{w}$ clearly has the units of feet, but when used in an energy sense it is better to express it in the form ft.lb./lb. of fluid.

In mechanics we have usually considered only two forms of energy, viz. Potential and Kinetic, but in hydrodynamics it is more convenient to consider pressure energy as distinct from potential energy in spite of their obvious interdependence as shown above.

For example, consider the quantity of fluid weight W moved from the surface to the point (2).

At (2) its total energy referred to the datum plane is made up of

Potential Energy Wh_2 and Pressure Energy $W\dfrac{p_2}{w}$.

Hence   Total Energy = Potential Energy + Pressure Energy

$$= Wh_2 \times W\frac{p_2}{w}.$$

But $$\frac{p_2}{w} = h,$$

$$\therefore \text{ Total Energy} = Wh_2 + Wh$$
$$= W(h_2 + h)$$
$$= Wh_1.$$

This represents its original potential energy and thereby illustrates the connection between potential and pressure energy.

We have here considered a fluid at rest, but, as will now be shown, providing kinetic energy is taken into account, it will apply equally as well to fluids in motion.

85. Bernoulli's Equation.—Again from the principle of the Conservation of Energy we know that the total energy in any system remains constant, although it may appear in a number of different forms. The transformation of potential energy into pressure energy has already been discussed, whilst for a case involving fluid motion the opening of a domestic water-tap affords a simple example of the conversion of town mains-pressure energy into kinetic energy.

Now, unfortunately, energy can also be dissipated in the form of friction, and whenever the *flow* of a fluid occurs the conversion of energy from one form to another always results in some loss of energy. However, the assessment of this energy loss is often rather involved and is usually based on laboratory investigation.

For this reason it is difficult to deduce strictly valid equations for the flow of actual fluids, whereas equations based initially on the flow of the *ideal* fluid can be made to yield results sufficiently accurate for our purposes.

We shall, therefore, first consider the case of a frictionless fluid flowing in the pipe shown in Fig. 238.

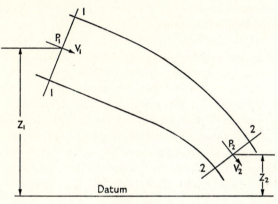

FIG. 238

The flow conditions on which the following energy equation is based must be clearly recognized ; they are stated as follows :

1. the pressure and velocity across any section are both considered uniform,
2. the rate at which fluid leaves section (2) is equal to the rate at which it enters section (1),

    i.e. the equation of continuity obtains,

or $wA_1V_1 = wA_2V_2 = W$ the weight rate of flow (w is constant).

Then the Principle of the Conservation of Energy provides that, without loss, the total energy at all sections is constant and at any section it is equal to the sum of the Pressure Energy, Kinetic Energy and Potential Energy.

Hence Total Energy at (1) = Total Energy at (2) = A Constant,

or $W\left(\dfrac{p_1}{w} + \dfrac{V_1^2}{2g} + Z_1\right) = W\left(\dfrac{p_2}{w} + \dfrac{V_2^2}{2g} + Z_2\right) = \text{Total Energy in ft.lb.}$

Eliminating W from both sides of the equation will give the total energy in ft.lb. per lb. of fluid, i.e. we obtain :

$$\frac{p_1}{w} + \frac{V_1{}^2}{2g} + Z_1 = \frac{p_2}{w} + \frac{V_2{}^2}{2g} + Z_2. \qquad . \qquad (1)$$

$$\underbrace{}_{\substack{\text{Pressure}\\\text{Head}}} \quad \underbrace{}_{\substack{\text{Velocity}\\\text{Head}}} \quad \underbrace{}_{\substack{\text{Potential}\\\text{Head}}}$$

and the energy equation in this form is known as *Bernoulli's* Equation for the perfect fluid.

Now real fluids differ considerably from the perfect fluid considered in the establishment of Bernoulli's Equation, and our success in estimating energy changes will largely depend on how we account for these differences in our modifications of Bernoulli's Equation. However, at this stage, it will be sufficient if we consider only one important modification—namely, loss of energy. From what has already been said it will be appreciated that a fluid cannot move, say, from section (1) to section (2) without some loss of energy—if only due to friction at the inner pipe surface. It should be evident then that there will be a loss of energy in the direction of motion—a fact which may be allowed for by writing equation (1) in the form :

Total Energy at (1) = Total Energy at (2)
+ Loss of Energy occurring between (1) and (2).

The Energy equation based on Bernoulli can now be used to express energy changes for a real fluid as follows :

$$\frac{p_1}{w} + \frac{V_1{}^2}{2g} + Z_1 = \frac{p_2}{w} + \frac{V_2{}^2}{2g} + Z_2 + \text{Losses (between (1) and (2))} \quad (2)$$

The loss of energy term in this equation is often expressed in the form $\dfrac{KV^2}{2g}$, in which K is an energy coefficient and $\dfrac{V^2}{2g}$ a relevant velocity head (see example 94, p. 313).

Even allowing for losses, equation (2) is still an over-simplification of the conditions found practically. From the introduction to this chapter sufficient has been said to indicate that the main assumptions of equation (1), namely, the existence of uniform velocity and pressure, can hardly be expected in actual fluid motion. Under practical conditions, therefore, we must content ourselves with calculations for kinetic energy based on the *average* flow velocity and also the knowledge that even if the pressure is not uniform the sum of $\frac{p}{w}$ and Z is likely to be constant at any section in a straight pipe. In a curved pipe we should expect centripetal forces to be developed (Chapter IV) and although Bernoulli's Equation does not allow for flow in a curved path, this effect must not be overlooked when considering the limitations of this equation.

Example 92

Water is discharged from a constant head tank at a point 60 feet below the tank level. If the diameter of the pipe at outlet is 2 inches, determine the discharge. Find also the pressure and velocity at a point 20 feet below tank level at which the pipe diameter is 4 inches.

Neglect all losses and take atmospheric pressure as 34 feet water.

Solution

The position of the datum is quite arbitrary and it is, therefore, chosen to coincide with the outlet point C. In this way the elevation or potential head Z_c is zero.

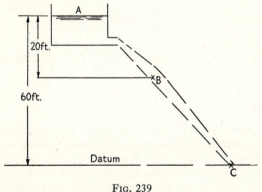

FIG. 239

Apply Bernoulli between A and C

$$\frac{p_A}{w} + \frac{V_A^{\,2}}{2g} + Z_A = \frac{p_c}{w} + \frac{V_c^{\,2}}{2g} + Z_c.$$

The water at A may be regarded as at rest, i.e. $V_A = 0$.

Hence $34 + 0 + 60 = 34 + \dfrac{V_c^2}{2g} + 0$

Hence $\dfrac{V_c^2}{2g} = 60$ and $V_c = $ **19·7** ft./sec.

Discharge $Q = A_c V_c = 19·7 \times \dfrac{\pi}{4}\left(\dfrac{2}{12}\right)^2 = $ **0·43 cusec.**

The velocity at B, 20 feet below tank level, can be obtained from the continuity equation $A_c V_c = A_B V_B$

or $V_B = V_c \cdot \dfrac{\pi 2^2}{\pi 4^2} = $ **4·92 ft./sec.**

To find the pressure at B the datum may either be taken through B or C.

Apply Bernoulli between A and B
    Datum through C.

$$\dfrac{p_A}{w} + \dfrac{V_A^2}{Zg} + Z_A = \dfrac{p_B}{w} + \dfrac{V_B^2}{2g} + Z_B$$

$$34 + 0 + 60 = \dfrac{p_B}{w} + \dfrac{(4·92)^2}{64·4} + 40$$

$$\dfrac{p_B}{w} = \textbf{53·625} \text{ ft. water.}$$

Alternatively $p_B = \dfrac{53·625 \times 62·4}{144} = $ **23·2 lb./in.²** (abs.)

Example 93

    Oil, of specific gravity 0·85, flows along a diverging tube AB at the rate of 200 cubic feet per second. The areas of cross-section at A and B are 4 sq. ft. and 10 sq. ft. respectively, and the centroid of the section at A is 20 feet above the centroid of the section at B. If the pressure at A is 20 lb./in.² abs., determine the total energy content of the oil at A in ft. lb./lb., taking the centroid of B as datum. Hence determine the pressure at B in lb./in.² abs., neglecting all losses.

Solution—Fig. 240

    The only point to watch in this question is that w, the specific wt., now becomes $(0·85 \times 62·4)$ instead of 62·4 lb./ft.³

From
$$Q = A_A V_A = A_B V_B$$

$$V_A = \frac{200}{4} = 50 \text{ ft./sec.} \quad V_B = \frac{100}{10} = 20 \text{ ft./sec.}$$

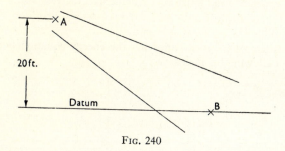

FIG. 240

Total energy at A with reference to datum through B :

$$= \frac{p_A}{w} + \frac{V_A^2}{2g} + Z_A$$

$$= \frac{20 \times 144}{0 \cdot 85 \times 62 \cdot 4} + \frac{(50)^2}{64.4} + 20.$$

Total energy = **113·4 ft.lb./lb.**

To determine the pressure at B we have (for no losses),

Total energy at A = Total energy at B,

i.e.
$$113 \cdot 4 = \frac{p_B}{w} + \frac{V_B^2}{2g} + Z_B,$$

or
$$113 \cdot 4 = \frac{p_B \times 144}{62 \cdot 4 \times 0 \cdot 85} + \frac{(20)^2}{64 \cdot 4} + 0$$

Hence
$$p_B = \textbf{39·4 lb./in.}^2$$

Example 94

It is required to discharge water to atmosphere at the rate of 1000 gallons per minute from a reservoir through a short length of 10-inch pipe followed by 1500 feet of 6-inch horizontal pipe. The loss of head in the 10-inch pipe may be neglected and that in the 6-inch pipe taken as $0 \cdot 04 \dfrac{V_2^2}{2g}$ per foot run of pipe. The exit loss from the reservoir is

A.M.—X

$0.5 \dfrac{V_1^2}{2g}$ and the contraction loss at the join of the two pipes $0.28 \dfrac{V_2^2}{2g}$, where V_1 and V_2 represent the velocities in the 10-inch and 6-inch pipes respectively.

What constant head must be maintained in the reservoir ?

Solution—Fig. 241

This is simply a case of applying the energy equation between A and B and allowing for losses.

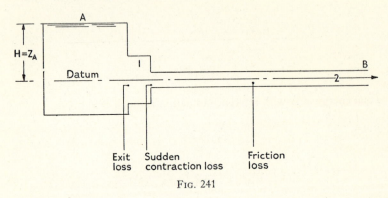

FIG. 241

The datum is taken through the pipe and Z_A then represents the required constant head in the reservoir.

Hence
$$\frac{p_A}{w} + \frac{V_A^2}{2g} + Z_A = \frac{p_B}{w} + \frac{V_B^2}{2g} + Z_B + \text{Losses (A to B)}$$

$$34 + 0 + Z_A = 34 + \frac{V_B^2}{2g} + 0 + \text{Losses},$$

or, in other words, the head in the reservoir must supply all the losses and the kinetic energy at discharge. (This is sometimes also termed a loss.)

$$Q = \frac{1000 \times 10}{60 \times 62 \cdot 4} = 2 \cdot 68 \text{ cusec.}$$

$$V_1 = \frac{26 \cdot 8}{\dfrac{\pi}{4}\left(\dfrac{10}{12}\right)^2} = 4 \cdot 9 \text{ ft./sec.} \qquad V_2 = V_B = \frac{2 \cdot 68}{\dfrac{\pi}{4}\left(\dfrac{6}{12}\right)^2} = 13 \cdot 6 \text{ ft./sec.}$$

Losses

$$\text{Exit loss} = 0 \cdot 5 \, \frac{V_1^2}{2g} = 0 \cdot 5 \times \frac{(4 \cdot 9)^2}{64 \cdot 4} = 0 \cdot 18 \text{ ft.}$$

Contraction Loss $=0.28\dfrac{V_2{}^2}{2g}=0.28\times\dfrac{(13.6)^2}{64.4}=0.80$ ft.

Friction Loss in 6-inch pipe $=0.04\dfrac{V_2{}^2}{2g}\Big/$ ft. $=\dfrac{0.04}{64.4}\times(13.6)^2\times1500=$ **171.6 ft.**

Total Loss   172.58 ft.

Now
$$Z_A=H=\frac{V_B{}^2}{2g}+\text{Losses}$$

$$=\frac{(13.6)^2}{64.4}+172.58$$

or   **H $=175.4$ feet.**

N.B.—Losses such as those which occur at exit from a reservoir and sudden contraction are referred to as minor losses when compared with friction losses in long pipe lengths. The above example illustrates the type of calculation in which these minor losses can quite safely be neglected.

Example 95

A test is made on a centrifugal pump which has a 6-inch diameter suction pipe and 4-inch diameter delivery pipe. When the pump is delivering 500 gallons of water per minute, the reading of the gauge on the suction pipe centre line is 18·5 inches mercury vacuum whilst the delivery pressure gauge reads 35 lb./in.² The centre of this gauge is 2 feet above the suction centre line. What power is the pump supplying to the water at this discharge ?

Solution—Fig. 242

There are two points to be observed in this question. (i) The suction gauge gives the vacuum at the centre of the suction pipe whilst

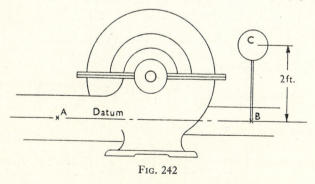

Fig. 242

the pressure gauge reading of 35 lb./in.² is the pressure at the centre line of the gauge. Both readings must be converted to feet of water at a common datum. (ii) The energy at B will be greater than that at A by the amount *added* by the pump. Hence the Energy equation now becomes :

$$\frac{p_A}{w} + \frac{V_A{}^2}{2g} + Z_A = \frac{p_B}{w} + \frac{V_B{}^2}{2g} + Z_B - \text{Energy supplied by pump,}$$

or,   Energy supplied by pump = Energy at B − Energy at A.

Pressure Head at A (absolute)

18·5″ Hg Vacuum is equivalent to $\dfrac{18\cdot5 \times 13\cdot6}{12} = 21$ ft. water.

This is head below atmospheric, hence

suction head = 34 − 21 = 13 ft. water (abs.)

Pressure Head at B

Pressure at gauge centre

$$p_c = 35 \text{ lb./in.}^2 \quad \frac{p_c}{w} = \frac{35 \times 144}{62\cdot4} = 81 \text{ ft. water (gauge).}$$

Equivalent head at point B, $\dfrac{p_B}{w} = \dfrac{p_c}{w} + Z_c = 81 + 2 = \textbf{83}$ ft. gauge.

Hence   pressure head = 34 + 83 = 117 ft. (abs.).

Velocities at A and B

By continuity $Q = A_A V_A = A_B V_B$

$$\frac{500 \times 10}{60 \times 62\cdot4} = \frac{\pi}{4}\left(\frac{6}{12}\right)^2 V_A = \frac{\pi}{4}\left(\frac{4}{12}\right)^2 V_B.$$

Hence $V_A = \textbf{6·8 ft./sec.}$ $V_B = \textbf{15·3 ft./sec.}$

Then by Energy equation,

$$\text{Pump head} = \left(\frac{p_B}{w} + \frac{V_B{}^2}{2g} + Z_B\right) - \left(\frac{p_A}{w} + \frac{V_A{}^2}{2g} + Z_A\right).$$

Taking datum through A, $Z_A = Z_B = \text{Zero.}$

$$\text{Pump head} = \left[117 + \frac{(15\cdot3)^2}{64\cdot4}\right] - \left[13 + \frac{(6\cdot8)^2}{64\cdot4}\right] = \textbf{106·9 ft. water.}$$

$$\text{Power supplied by pump} = \frac{\text{Head (ft.)} \times \text{Quantity (lb./min.)}}{33000}$$

$$= \frac{106 \cdot 9 \times 500 \times 10}{33000}.$$

$$\text{Power} = \mathbf{16 \cdot 2 \text{ H.P.}}$$

The remainder of this chapter is devoted to a very elementary treatment of three types of instrument in order to illustrate the application of the energy equation to flow measurement.

86. The Orifice.—An orifice is simply a hole, usually in a plate on the side of a tank, and is often sharp-edged. Its use as a flow-measuring device can be demonstrated by considering the large tank in Fig. 243, in which is shown fluid being discharged to atmosphere under a constant head h above the orifice centre-line.

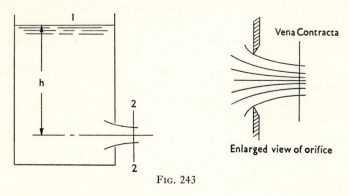

Vena Contracta

Enlarged view of orifice

Fig. 243

In order to calculate the quantity flowing through the orifice we require to know the velocity of flow.

The connection between velocity and head can be obtained from the energy equation by considering the points (1) and (2) at the fluid surface and near the orifice outlet.

Since the fluid approaches and passes through the orifice in a curved path we cannot apply Bernoulli when section (2) is taken in the actual plane of the orifice (see page 311). However, the jet is found to converge to a minimum area at a distance roughly half the orifice diameter away from the face of the orifice. This minimum or contracted area is called the *vena contracta* and is the first section over which the velocity is almost wholly axial. Hence the reason for choosing section (2) at the position shown.

If the tank is large, the velocity at the surface is negligible and the pressure at (1) and (2) is atmospheric.

Applying the Energy equation to (1) and (2),

$$\frac{p_1}{w} + \frac{V_1{}^2}{2g} + Z_1 = \frac{p_2}{w} + \frac{V_2{}^2}{2g} + Z_2 + \text{Losses (1-2)}$$

$$34 + 0 + h = 34 + \frac{V_2{}^2}{2g} + 0 + \text{Losses}.$$

Then
$$h = \frac{V_2{}^2}{2g} + \text{Losses}. \qquad . \qquad . \qquad . \qquad (1)$$

If there were no losses then clearly $h = \frac{V_2{}^2}{2g}$ and $V_2 = \sqrt{2gh}$, which may be recognised as the velocity attained by a solid body falling freely from a height h. For no energy loss, then, V_2 can be looked upon as the ideal jet velocity. However, as suggested by equation (1) V_2 will actually be slightly less than $\sqrt{2gh}$ on account of friction. This friction is minimised by making the orifice sharp-edged on the up-stream side as shown in Fig. 243, but cannot be entirely eliminated.

The ratio, actual to ideal velocity at the *vena contracta*, is called the coefficient of velocity and has the symbol C_v,

$$\text{i.e. } C_v = \frac{\text{Actual Velocity}}{\sqrt{2gh}} \text{ or } V_2 = C_v\sqrt{2gh}.$$

The Quantity follows from Area × Velocity.

Hence if A_2 is the area of flow at the *vena contracta*, we have

$$Q = A_2 C_v \sqrt{2gh}. \qquad . \qquad . \qquad . \qquad (2)$$

Now the ratio of the area of the jet A_2 to the actual orifice area A is called the Coefficient of Contraction, C_c.

Therefore
$$C_c = \frac{A_2}{A} \text{ and } A_2 = AC_c.$$

Substituting this value in equation (2) gives,

$$Q = C_c C_v \times A \sqrt{2gh}.$$

On account of the difficulty found in measuring C_c and C_v it is more convenient to replace the product C_cC_v by a single coefficient C_d called the Coefficient of Discharge.

Then finally,

$$\text{Actual } Q = C_dA\sqrt{2gh}. \qquad . \qquad . \qquad (3)$$

Clearly C_d is the ratio of the actual discharge to the ideal one based on no contraction or loss of energy.

The value of C_d can be determined for a given orifice and head by weighing the discharge over a suitable period, the value so obtained can then be used to determine discharges under other values of head.*

Fig. 244 shows an alternative arrangement of a sharp-edged orifice as used in pipe-flow measurements.

In this case the value of h in equation (3) becomes the difference in head across the orifice between a suitably placed upstream tapping and a downstream tapping made to coincide approximately with the *vena contracta* as in Fig. 244.

Equation (3) still holds for this arrangement.

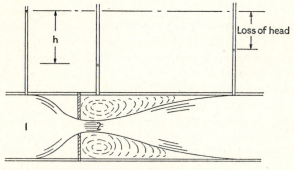

Fig. 244

Example 96

When water is discharged through a 2-inch diameter sharp-edged orifice under a head of 15 feet the measured rate of flow is found to be 453 gallons in 3 minutes. Determine the coefficient of discharge and the probable discharge when the head over the orifice is 20 feet.

* Actually C_d is not constant, particularly at low heads. It varies with the ratio of head to orifice diameter. This effect is considered further in later studies, but a commonly accepted value is in the region of 0·62.

Solution

$$Q = C_d A \sqrt{2gh}.$$

Q is in cusec, then discharge/min. $= \dfrac{453}{3} = 151$ gal./min.

$$\therefore \quad Q = \frac{151 \times 10}{62 \cdot 4 \times 60} = 0 \cdot 404 \text{ cusec.}$$

$$A = \frac{\pi}{4}\left(\frac{2}{12}\right)^2 = 0 \cdot 0218 \text{ ft.}^2$$

$$C_d = \frac{Q}{A\sqrt{2gh}} = \frac{0 \cdot 404}{0 \cdot 0218\sqrt{15 \times 64 \cdot 4}} = \mathbf{0 \cdot 597}$$

On the assumption that C_d remains constant over the given head range Q is proportional to $\sqrt{h}$.

Hence　　Quantity at 20-ft. head $= 151\sqrt{\dfrac{20}{15}} = \mathbf{174 \cdot 5 \ gal./min.}$

87. The Venturi Meter.

87. The Venturi Meter.—Whichever instrument is used for flow measurement some loss of energy is bound to occur, and from this point of view the pipe orifice of Fig. 244 is not entirely satisfactory. It certainly has the advantages of being cheap and relatively easy to instal, but in cases where such factors are unimportant it is much better to use a Venturi meter.

This instrument is usually made as a permanent installation in the pipe-work and consists simply of a tube which converges to a minimum area called the throat and then diverges gradually back to the normal pipe diameter. Its operation, therefore, is like that of the orifice of Fig. 244 and is based on the conversion of pressure energy in the pipe to kinetic energy in the throat. However, the loss of energy due to turbulence, which occurs just downstream in the case of the orifice, is largely avoided in the Venturi meter by means of the *gradual* divergence beyond the throat. Hence the overriding advantage of this meter compared with the pipe orifice is that it can be so manufactured as to incur almost negligible loss of energy. Compare Figs. 244 and 245.

Neglecting losses, then, from Fig. 245, Bernoulli's Equation gives :

$$\frac{p_1}{w} + \frac{V_1^2}{2g} + Z_1 = \frac{p_2}{w} + \frac{V_2^2}{2g} + Z_2.$$

For a horizontal meter $Z_1 = Z_2 = $ zero.

Now $\dfrac{p_1}{w} - \dfrac{p_2}{w} = h,$

$$\therefore\ h = \frac{V_2{}^2}{2g} - \frac{V_1{}^2}{2g}. \qquad \cdot \qquad \cdot \qquad \cdot \quad (1)$$

Then from continuity $A_2V_2 = A_1V_1$ and $V_2 = \dfrac{A_1}{A_2}V_1.$

Substituting for V_2 in equation (1) :

$$h = \left(\frac{A_1}{A_2}\right)^2 \frac{V_1{}^2}{2g} - \frac{V_1{}^2}{2g}$$

or $$h = \frac{V_1{}^2}{2g}\left[\left(\frac{A_1}{A_2}\right)^2 - 1\right].$$

If $\dfrac{A_1}{A_2} = n = $ Constant for the meter

$$h = \frac{V_1{}^2}{2g}[n^2 - 1].$$

Transposing $$V_1 = \sqrt{\frac{2gh}{n^2 - 1}}.$$

Therefore $$Q = A_1V_1 = A_1\sqrt{\frac{2gh}{(n^2 - 1)}}, \qquad \cdot \qquad \cdot \quad (2)$$

or $$Q = A_1\sqrt{\frac{2g\left(\dfrac{p_1 - p_2}{w}\right)}{(n^2 - 1)}}.$$

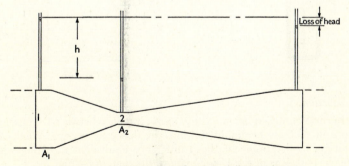

'h' is the head difference in ft. of the following liquid.

Fig. 245

Since some slight loss of energy is unavoidable the quantity will be less than that given by equation (2). Introducing a discharge coefficient C_d, we have the final equation

$$Q = C_d A_1 \sqrt{\frac{2gh}{(n^2 - 1)}}.$$

Now for any one meter $\sqrt{2g}$, A_1 and $(n^2 - 1)$ are constant, and since C_d is reasonably so this equation is often given in the form $Q = K\sqrt{h}$, in which K is called the meter coefficient. C_d is of the order of 0·98.

Example 97

A Venturi meter measures the flow of water in a 3-inch diameter pipe. The difference of head between the entrance and the throat of the meter is measured by a U-tube containing mercury, the space above the mercury on each side being filled with water. What should be the diameter of the throat of the meter in order that the difference of the levels of the mercury shall be 10 inches when the quantity of water flowing in the pipe is 140 gal./minute? Assume the discharge coefficient as 0·97.

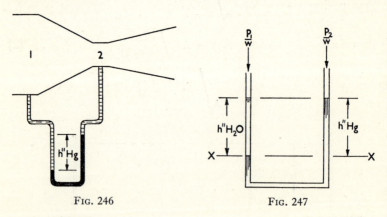

FIG. 246   FIG. 247

Introduction

In the Venturi equation h is the head difference between the tapping points in feet of the liquid flowing—in this case water. We have, therefore, to determine the head difference equivalent to h'' Hg with water in both legs of the U-tube, above the mercury—Fig. 246.

Consider the enlarged U-tube in Fig. 247 showing the heads on the common surface XX.

We have, in the left-hand limb :

$$\text{head at XX} = \frac{p_1}{w} + \text{head due to } h \text{ inches water.}$$

In the right-hand limb :

$$\text{head at XX} = \frac{p_2}{w} + \text{head due to } h \text{ inches Hg.}$$

The heads in these two limbs are equal, and expressing them in feet of water we get :

$$\frac{p_1}{w} + \frac{h''}{12} = \frac{p_2}{w} + \frac{h'' \times 13 \cdot 6}{12},$$

or

$$\frac{p_1 - p_2}{w} = \frac{12 \cdot 6h}{12} = 1 \cdot 05h,$$

where h is the length of mercury column.

Therefore, for the case of a mercury U-tube, we obtain :

$$Q = C_d A_1 \sqrt{\frac{\left(\frac{p_1 - p_2}{w}\right) 2g}{(n^2 - 1)}} \text{ or } Q = C_d A_1 \sqrt{\frac{1 \cdot 05h \cdot 2g}{(n^2 - 1)}}.$$

Solution

The only unknown in the Venturi equation is n^2, which is the ratio of areas. This will give us the required throat diameter.

$$\frac{p_1}{w} - \frac{p_2}{w} = 1 \cdot 05h = 1 \cdot 05 \times 10 = \textbf{10·5 ft.} \text{ water}$$

$$A_1 = \frac{\pi}{4}\left(\frac{3}{12}\right)^2 = \frac{\pi}{64} \text{ ft.}^2$$

$$Q = \frac{140 \times 10}{62 \cdot 4 \times 60} = \textbf{0·374 cusec.}$$

Then

$$0 \cdot 374 = 0 \cdot 97 \frac{\pi}{64} \sqrt{\frac{10 \cdot 5 \times 64 \cdot 4}{(n^2 - 1)}}.$$

Transposing gives

$$n = 3 \cdot 45.$$

Now

$$n = \frac{A_1}{A_2} = \frac{\pi d_1{}^2}{\pi d_2{}^2},$$

i.e. $3 \cdot 45 = \dfrac{3^2}{d_2{}^2}$ or $d_2 = \textbf{1·62 inches.}$

Example 98

A Venturi meter, with 18-inch throat, is fitted into a 36-inch diameter pipe. It is placed with the axis of the meter inclined at 45° to the horizontal. The head difference across the meter is measured by a mercury manometer. Calculate the discharge in gallons of water per minute when the manometer reads 7 inches.

Introduction

Consider the inclined meter in Fig. 248.

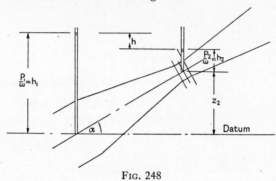

FIG. 248

Applying Bernoulli :

$$\frac{p_1}{w} + \frac{V_1{}^2}{2g} = \frac{p_2}{w} + \frac{V_2{}^2}{2g} + Z_2$$

or

$$\frac{V_2{}^2 - V_1{}^2}{2g} = \frac{p_1}{w} - \frac{p_2}{w} - Z_2.$$

But

$$\frac{p_1}{w} = h_1 \text{ and } \frac{p_2}{w} = h_2,$$

$$\therefore \frac{p_1}{w} - \frac{p_2}{w} - Z_2 = h_1 - h_2 - Z_2,$$

which from Fig. 248 is clearly equal to h

Hence

$$\frac{V_2{}^2 - V_1{}^2}{2g} = h,$$

which is the same as equation (1) for a horizontal meter. (See page 321.)

Therefore, the inclination of the meter has no effect on the discharge equation, provided the pressure difference is measured as shown above.

Solution

Assuming $C_d = 1$. Then $Q = A_1 \sqrt{\dfrac{2g \, 1 \cdot 05 h}{(n^2 - 1)}}$

$$A_1 = \frac{\pi 3^2}{4} = \frac{9\pi}{4}, \quad n^2 = \left(\frac{3}{1 \cdot 5}\right)^2 = 4.$$

Then
$$Q = \frac{9\pi}{4} \sqrt{\frac{64 \cdot 4 \times 1 \cdot 05 \times 7}{4 - 1}}$$

$$= \textbf{89·2 cusec.}$$

Discharge in gal./min. $= \dfrac{89 \cdot 2 \times 62 \cdot 4 \times 60}{10}$

Discharge $= \textbf{33400 gallons/min.}$

88. Rectangular and Triangular Notches.—If the head above the orifice in Fig. 243 is reduced so that the orifice does not run full the orifice is then called a notch.

Fig. 249 illustrates a notch in a channel having a relatively low flow.

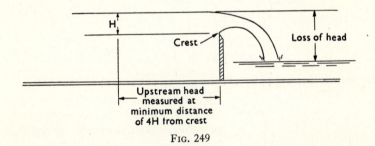

FIG. 249

At some distance upstream, for the simple case illustrated, the water may be considered at rest, but, as will be observed, in passing downstream over the crest the water surface drops below this undisturbed level. Because of this it is not found practical to measure the head causing flow at a point coinciding with the plane of the notch. Instead, the head is measured some distance upstream as shown in Fig. 249.

(i) *The Rectangular Notch*

An expression for the quantity passing over a rectangular notch can be obtained by assuming the flow over the crest to take place parabolically as in Fig. 250.

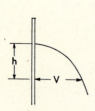
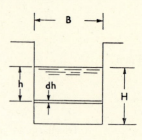

Fig. 250

At an elemental area Bdh at h below the upstream depth the velocity

$$V = \sqrt{2gh}.$$

Then discharge through the elemental area

$$dq = \sqrt{2gh} \,.\, B \,.\, dh.$$

By integration for the total quantity

over the notch.

$$Q = B\sqrt{2g}\int_{0}^{H} \sqrt{h} \,.\, dh,$$

$$\therefore \;\; Q = \tfrac{2}{3}B\sqrt{2g} \,.\, H^{3/2}.$$

In actual fact there is a contraction of flow over the notch as suggested in Fig. 251.

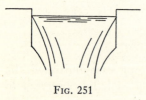

This effect and the inevitable presence of friction can be allowed for by introducing a coefficient of discharge into the theoretical equation. The working notch equation (*) becomes

Fig. 251

$$Q = \tfrac{2}{3}C_d B \sqrt{2g} H^{3/2}.$$

* If the velocity where H is measured is appreciable the total energy upstream becomes $H + \dfrac{V_0{}^2}{2g}$, where V_0 is called the velocity of approach. This modifies the limits of integration, but since V_0 is difficult to assess accurately and its effect is usually small for low flows it is omitted from the above equation.

(ii) *The Triangular Notch*

A similar equation may be obtained for the triangular notch shown in Fig. 252.

Let the vertex angle be θ.

Velocity at depth $h = \sqrt{2gh}$.

By similar triangles the length of the elemental strip

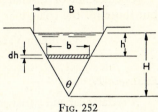

FIG. 252

$$b = \frac{B}{H}(H - h).$$

Then elemental area $= \frac{B}{H}(H - h)dh$,

$$\therefore dq = \sqrt{2gh} \cdot \frac{B}{H}(H - h)dh.$$

Integrating

$$Q = \sqrt{2g}\frac{B}{H}\int_0^H (Hh^{1/2} - h^{3/2})dh$$

$$= \sqrt{2g}\frac{B}{H}\left[\frac{2}{3}H^{5/2} - \frac{2}{5}H^{5/2}\right],$$

$$\therefore Q = \sqrt{2g}\frac{B}{H} \cdot \frac{4}{15} \cdot H^{5/2}.$$

Substituting

$$B = 2H \tan \frac{\theta}{2},$$

we get Theoretical

$$Q = \frac{8}{15}\sqrt{2g} \tan \frac{\theta}{2} H^{5/2},$$

and introducing a coefficient of discharge,

$$\text{Actual } Q = \frac{8}{15}C_d \sqrt{2g} \tan \frac{\theta}{2}H^{5/2}.$$

For notches with 90° vertex angle $\tan \frac{\theta}{2} = $ unity.

Notches Compared

The triangular notch is essentially a small quantity meter but has the advantage of being able to accommodate a wider head range with more accuracy than the rectangular notch. The rectangular notch is used for relatively larger quantities at high heads. It is unreliable at low heads.

Coefficients of discharge for both notches are not constant but those for the triangular show the least variation with head.

Common Values for C_d. 0·59 – 0·62.

Example 99

A rectangular and a triangular notch are used in parallel to measure the discharge during a test on a centrifugal pump. The rectangular notch is 10 inches wide, the head across it being 3·60 inches and the head across the 90° triangular notch is 4·75 inches. Assuming that the coefficient of discharge equals 0·60 for both notches, calculate the pump discharge in gallons per minute.

Solution

For rectangular notch :

$$Q = \frac{2}{3} \, C_d B \sqrt{2g} H^{3/2} \text{ cusec}$$

$$= \frac{2}{3} \times 0·6 \times \frac{10}{12} \sqrt{2g} \left(\frac{3·6}{12}\right)^{3/2}$$

$$= 2·68 \times 0·163$$

$$Q = \textbf{0·436 cusec.}$$

For triangular notch :

$$= \frac{8}{15} C_d \, \tan \frac{\theta}{2} \sqrt{2g} H^{5/2}$$

$$= \frac{8}{15} \times 0·6 \times 1 \times \sqrt{2g} \left(\frac{4·75}{12}\right)^{5/2}$$

$$= 2·57 \times 0·0985$$

$$= \textbf{0·253 cusec.}$$

$$\text{Total Quantity} = 0·436 + 0·253$$

$$= 0·689 \text{ cusec.}$$

Hence Quantity discharged in gallons per min. $= 0·689 \times 6·24 \times 60$

$$= \textbf{257·7}$$

EXERCISES XI

1. Water flows along a pipe from a point A to a point B at the rate of 300 cu.ft./min. At A the pipe diameter is 8 inches and the pressure is 40 lb./in.² abs. At B, which is 25 feet below A, the pipe diameter is 5 inches. Determine :

(*a*) the total energy of the water at A in ft.lb. per lb., assuming the datum level to be the horizontal through B ;

(*b*) the pressure at B in lb./in.²

Neglect fluid friction and take the specific weight of water as 62·4 lb./ft.³

2. State Bernoulli's theorem giving the units in which each quantity is expressed.

Water is pumped 'uphill' in a pipe inclined at 30° to the horizontal. At a point A the internal diameter of the pipe is 6 inches, and at a higher point, B, 6 feet from A, the internal diameter is 3 inches—the change being uniform. If the pressure at A is 60 lb./in.², find the pressure at B when the flow is at the rate of 1 cu.ft./sec.

3. A horizontal pipe-line of diameter D conveys water at a steady rate. It is provided with a fitting—a Venturi meter—in which the bore contracts to a throat diameter *d*, and thereafter tapers to the original diameter D. The drop in pressure between the entry and throat of the meter is equivalent to H feet of head. Obtain an expression for the flow in cu.ft. per sec. Friction may be neglected.

In such an arrangement the pipe diameter is 6 inches and the throat diameter is 2 inches. When the flow is 0·5 cu.ft. per sec. the pressure at entry is 10 lb./in.² Determine the pressure at the throat in the same units.

(Weight of water, 62·2 lb./cu.ft.)   I.Mech.E.

4. In an experiment to determine the coefficients of contraction, velocity and discharge for a circular orifice the following readings were taken :

Diameter of orifice = 0·35 in.

Head of water above centre of orifice = 4 ft.

Discharge of water 200 lb. in 469 sec.

The jet discharged into a tank in which the water surface was 6·1 inches below the centre line of the orifice. The jet struck the water surface at a horizontal distance of 33·5 inches from the plane of the orifice. Determine the values of the coefficients for this orifice.

N.C.T.E.C.

5. A horizontal Venturi meter has an entry diameter of 9 inches and a throat diameter of 3 inches. The coefficient for the meter is 0·96.

A.M.—Y

The difference in pressure between the entrance and throat is equivalent to 2 inches of mercury. Calculate the quantity of water flowing in gallons per min.

Take 1 cu.ft. of water to weigh 62·5 lb. and the specific gravity of mercury as 13·6.

Bernoulli's Equation may be assumed, but any other formulae used must be proved.

N.C.T.E.C.

6. A pipe conveying water rises at one place from a position A to a position B, 10 feet above A. At the lower level the pipe diameter is 12 inches and the pressure there remains constant at 5 lb./in.² gauge, even although the flow quantity may vary. At B a Venturi throat is formed so that flow changes may be observed by means of gauge readings there. The maximum flow is at the rate of 2500 gal. per min., and for this condition the pressure at the throat at B is not to be lower than 5 lb./in.² below atmosphere. Determine the minimum throat diameter permissible.

If the flow rate falls by 10 per cent below the maximum, estimate the change in the gauge reading at B.

(Specific weight of water, 62·5 lb./ft.³ ; 1 gallon of water weighs 10 lb.)

I.Mech.E.

7. Describe briefly, with the aid of a sketch, the principle of a Venturi meter.

A Venturi meter, with a throat diameter of 2 inches, is built into a 6-inch diameter horizontal water main. The difference in pressure between the throat and the main is measured by a mercury manometer and the difference in mercury levels in the two limbs is 3 inches.

Using Bernoulli's theorem, calculate the theoretical flow in gallons/hour. Assume the space above the mercury in the manometer to be filled with water.

(Water weighs 62·5 lb./ft.³ Specific gravity of mercury 13·6.)

8. A 1-inch diameter sharp-edged orifice discharges 175 gallons of water per minute when the head is 300 feet. The diameter of the narrowest portion of the jet is 0·8 inch. Calculate the coefficients of discharge, contraction and velocity.

(Water weighs 62·5 lb./ft.³)

9. A vessel containing water having a 1-inch diameter orifice in one side is suspended so that any displacing force can be accurately measured. The orifice is first plugged ; on removal of the plug the horizontal force required to keep the vessel in place, applied opposite the orifice, was 3·6 lb. By measurement the discharge was found to be 31 gallons/min. The level of the water in the vessel was maintained at a constant height of 9 feet above the orifice. Determine C_d, C_v and C_c.

10. State Bernoulli's theorem and describe, with the aid of a sketch, the principle of a Venturi meter.

A Venturi meter has a throat diameter of 3 inches and is fitted to a 6-inch water main. A mercury manometer fitted to the main and to the throat indicates a difference in head of 15 inches of mercury. If the meter has a coefficient of 0·98, calculate by working from first principles the discharge in gallons per minute.

(Specific gravity of mercury 13·6.)

11. The cooling water from a small condenser passes along a horizontal pipe of 3 inches diameter to a measuring tank, from which it is discharged through a sharp-edged orifice of $1\frac{1}{2}$ inches diameter. For the condition of maximum flow the head in the tank over the centre of the orifice is 5 feet. If the coefficient of discharge for the orifice is 0·62, determine the corresponding speed of flow in the pipe.

A Venturi meter is to be used instead of the measuring tank and one with a throat diameter of 1 inch is selected for fitting in the pipeline. Estimate the pressure drop between the entry and the throat of this meter in lb./in.^{2} for the same rate of flow.

(Specific weight of water, 62·2 lb./ft.^{3})   I.Mech.E.

12. Water flows along a pipe AB at the rate of 100 gal./min. The pipe is 2 inches diameter at A and 1·0 inch diameter at B, and the height of A above B is 5 feet. The pipe is horizontal at B, and on discharge from B the water strikes a fixed vertical plate. Determine the pressure in pounds per square inch at A and the force exerted by the jet on the fixed plate.

(Density of water is 62·4 lb./ft.^{3} and 1 gal. weighs 10 lb.)

U.L.C.I.

13. A rectangular notch discharges 425 gallons of water per minute when the head above the sill is 6 inches. Assuming the coefficient of discharge is 0·6, find the breadth of the notch. What is the discharge in gallons per minute when the head is increased to 2 feet? The coefficient of discharge remains the same.

APPENDIX

1. d'Alembert's Principle

An alternative approach to many problems in dynamics is to transform them into equivalent problems in statics.

This approach is due to d'Alembert and the principle involved in the method is now explained.

Newton's Second Law is expressed by the equation :

$$F = mf,$$

where F is the resultant force which acts on a body mass m to give it an acceleration equal to f.

This equation can also be written in the form :

$$F - mf = 0, \qquad . \qquad . \qquad . \qquad . \qquad (1)$$

which expresses the fact that the resultant force F minus the force represented by mf is zero.

The alternative form of Newton's Second Law expressed by equation (1) is similar to the force equilibrium equation employed in statics and it is on this fact that the method is based. Referring now to Fig. 253(a) let F be the resultant of all the external forces F_1, F_2, etc., which give to the body of mass m an acceleration f.

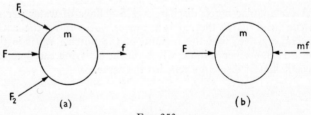

Fig. 253

Suppose now that an imaginary force equal to mf is applied to the body, in a direction opposite to the acceleration f, to give the condition shown in Fig. 253(b).

The body can now be thought of as in 'dynamic' equilibrium, as the summation of the forces acting on it is now zero. (Σ Forces $= 0$.)

The ordinary laws of statics can now be applied to the solution of the problem.

The resultant external force F, which is equal to *mf*, is called the 'effective' force.

The *imaginary* force which, when applied to the body, provides dynamical equilibrium is called the 'inertia' force, since it is proportional to the mass and hence the inertia of the body. As it is equal and opposite to the effective force F, this imaginary force is also known as the reversed effective force.

The use of d'Alembert's principle is largely a matter of personal preference, as there is no particular advantage over the method already suggested.

As an illustration of this alternative approach it will now be applied to the case of the conical pendulum given on page 95.

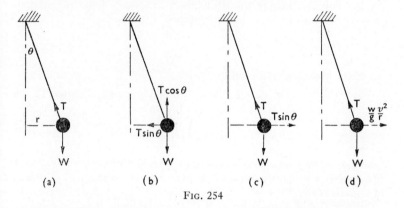

Fig. 254

For the given configuration of the pendulum in Fig. 254(a) the actual or effective forces acting ON the bob are the tension T and its weight W.

In Fig. 254(b), T is resolved into its components, and we now have the three effective forces acting on the bob, viz. $T \cos \theta$, $T \sin \theta$ and W.

Since $T \cos \theta$ and W provide vertical equilibrium, it is evident that the *imaginary* force which must be applied to the bob to give horizontal equilibrium is $T \sin \theta$ in a direction *outwards* from the centre of rotation—Fig. 254(c). The bob in Fig. 254(c) is now in dynamic equilibrium.

But we know that the horizontal component of T provides the centripetal force

$$\frac{W}{g}\frac{v^2}{r}.$$

Therefore, the imaginary—or reversed effective—force must also equal

$$\frac{W}{g}\frac{v^2}{r}.$$

Hence the final arrangement for solution by d'Alembert's principle is that given in Fig. 254(d), from which we have, by the equations of statics :

$$\text{For horizontal equilibrium} \quad \text{T} \sin \theta = \frac{\text{W}}{g} \frac{v^2}{r}$$

and for vertical equilibrium T cos θ = W.

These equations should be compared with those given on page 96.

In the conical pendulum and similar problems on rotating bodies the reversed effective force which is equal and opposite to the centripetal force is called the CENTRIFUGAL FORCE.

It must be stressed that the centrifugal force is not an actual force ON the bob but merely the d'Alembert imaginary force required to provide the normal dynamic configuration of the body.

2. Mass Moment of Inertia

In Chapter VIII on the theory of bending and Chapter IV on dynamics we encountered integrals of the form

$$\int ay^2 \text{ and } \int my^2.$$

The first of these integrals represents the second moment of area for a plane surface whilst the second represents the mass moment of inertia of a given body.

In view of the similarity of both these integrals it is the common practice to use the symbol I to represent both of them. For this reason the terms second moment of area and moment of inertia are often employed for the same integral $\int ay^2$. This is particularly so in the case of problems on bending of beams.

In order to avoid any possible confusion it is important that the integral $\int my^2$ be called the Mass Moment of Inertia and not just the moment of inertia, even though the context may clarify the meaning.

Hence

Second Moment of Area I $= \int ay^2$ (sometimes called moment of inertia)

Mass Moment of Inertia I $= \int my^2$.

The following theorems apply equally to second moments of area of plane areas.

3. Perpendicular Axis Theorem

Consider the body in Fig. 255.

The point O in the body is the intersection of the three perpendicular axes x, y and z—the z axis through O being perpendicular to the plane of the diagram (this axis is often referred to as the polar axis).

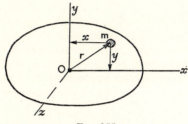

Fig. 255

For any element of mass m the mass moment of inertia :

$$I_x \text{ about the } x \text{ axis} = \int my^2,$$

$$I_y \text{ about the } y \text{ axis} = \int mx^2$$

and

$$I_z \text{ about the } z \text{ axis} = \int mr^2.$$

From the geometry of Fig. 255 we have $r^2 = x^2 + y^2$,

$$\therefore \ I_z = \int mr^2 = \int m(x^2 + y^2)$$

or

$$I_z = \int mx^2 + \int my^2.$$

Then

$$I_z = I_y + I_x.$$

Thus, the mass moment of inertia of any body with respect to an axis through any point O is equal to the sum of the mass moments of inertia with respect to any other two mutually perpendicular axes through the same point. This is called the perpendicular axis theorem.

Note on Second Moment of Area

When dealing with second moments of area we use the symbol J to denote the second moment of area about the polar axis. (See article 66.)

Then for a circular section it follows from the perpendicular axis theorem that if $I_{\text{DIA.}}$ represents the second moment of area about a diameter

$$J = 2I_{\text{DIA}},$$

i.e. $\dfrac{\pi d^4}{32} = 2 \cdot \dfrac{\pi d^4}{64}.$

4. Parallel Axis Theorem

When the mass moment of inertia of a body about an axis through the centre of gravity is known, the mass moment of inertia about any parallel axis can be calculated by means of the parallel axis theorem.

In Fig. 256 the perpendicular axes x and y meet at G, the centre of gravity of the body, and z is any axis parallel to x and distance h from x.

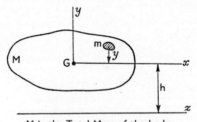

M is the Total Mass of the body.

Fig. 256

The elemental mass shown in the figure is at a distance $(y + h)$ from the Z axis.

Then
$$I_z = \int m(y + h)^2 = \int m(y^2 + 2yh + h^2),$$

$$\therefore \ I_z = \int my^2 + \int 2myh + \int mh^2.$$

Now
$$\int my^2 = I_x \text{ and } \int mh^2 = h^2 \int m = Mh^2.$$

The second integral $\int 2myh$ may be written as $2h \int my$ and $\int my$ is zero since it represents the first moment of mass about the axis x *through* G.

Therefore
$$I_z = I_x + Mh^2.$$

Hence the mass moment of inertia of a body about an axis parallel with an axis through its centre of gravity is equal to the mass moment of inertia of the body about its centre of gravity plus the mass multiplied by the square of the distance between the parallel axes.

ANSWERS TO EXERCISES

EXERCISES I : Page 26

1. $R_A = 1\frac{4}{7}$ tons. R_C $4\frac{3}{7}$ tons.
2. Anticlockwise couple of 2 ton ft.
3. 28·6 lb.
4. R_C 2·68 tons. R_B 22·3 tons at 35·6° to horizontal.
5. 7·1 tons at 40° 26′ to horizontal.
6. A : 28·85 lb. B : 15·45 lb. C.: 10·95 lb.
7. Clockwise couple of 8·6 lb. ft.
8. Rod at 30° to vertical. Force on peg 60 lb.
9. Hinge reaction 271 lb. at 5·1° to horizontal. Tension in rope 312 lb.
10. (a) 19·5 tons. (b) 15·3 tons at 35° 50′ to horizontal.
11. $R_A = 2\frac{1}{4}$ tons, R_E $1\frac{3}{4}$ tons. AB, BC, CD, DE, BG, GD. Force in CG = 1·45 tons.
12. (a) $R_1 = 5$ tons, $R_4 = 7$ tons; (b) $2 - 5 = 2\sqrt{3}$-tons Tie, $3 - 5 = 2/\sqrt{3}$-tons Tie.
13. (a) $R_A = 5\cdot38$ tons, $R_B = 2\cdot62$ tons.
    (b) Forces in tons (1) = 6·30 Strut, (2) = 5·70 Strut, (3) = 3·30 Strut, (4) = 5·30 Strut, (5) = 4·50 Tie, (6) = 4·50 Tie, (7) = 3·20 Tie, (8) = 0·68 Strut, (9) = 1·60 Tie, (10) = 1·96 Strut, (11) zero.
14. Forces in tons : (1) = 7·9 Strut, (2) = 3·5 Strut, (3) = 2·5 Strut, (4) = 0·9 Tie, (5) = 4·7 Tie, (6) = 7·4 Tie, (7) = 4·05 Tie, (8) = 2·05 Strut, (9) = 4·25 Tie.
15. (a) $R_A = 6/\sqrt{3}$ tons horizontal. $R_E = 4$ tons at 60° to the vertical.
    (b) Forces in tons : $DC = \dfrac{4}{\sqrt{3}}$ Strut, $BC = \dfrac{2}{\sqrt{3}}$ Tie, $BD = \dfrac{4}{\sqrt{3}}$ Tie, $ED = 4/\sqrt{3}$ Strut, $EB = 4/\sqrt{3}$ Strut, $AB = 6/\sqrt{3}$ Tie.
16. (1) = 9·9-tons Tie, (2) = 7-tons Tie, (3) = $\sqrt{2}$-tons Tie, (4) = 8-tons Strut.
17. Force in EI = 2·4-tons Tie ; Struts : AB, CJ, EH, FG, BC, CD, DE, EF ; Zero force in AJ, HG, DI.
18. BF = 6-tons Strut ; FG = 3·46-tons Tie ; GD = 3·46-tons Tie.
19. $R_A = 1\cdot5$ tons vertical, $R_B = 2\cdot75$ tons at 49° to horizontal. Force in (1) = 1·2-tons Strut, (2) = 2·6-tons Strut, (3) = 3·9-tons Tie.
20. Forces in lb. : BE = 650 Tie ; CF = zero. AE = 830 Strut, AG = 1250 Strut, DG = 250 Strut ; EF = 830 Strut ; FG = 1030 Tie. Y = 1000 lb. horizontal, Z = 1000 $\sqrt{2}$ lb. at 45° to wall.
21. Forces in tons : ED = 1·59 Tie ; DC = $1/\sqrt{2}$ Tie ; AD = 0·9 Strut ; AB = 0·5 Strut ; BC = 0·5 Strut ; BD = 0·25 Tie. $R_A = 1\cdot29$ tons at 62° to vertical.

22. Forces in tons : AD = 16 Tie ;   CD = 9 Tie ;   AB = 15 Tie ;
BC = 20 Strut ; AC = 5 Tie.
Components at C : 12 tons vertically upwards and 16 tons horizontally to the left.

EXERCISES II : Page 54

1. (i) 10 lb., (ii) 8·96 lb.
2. 200 ft.lb.
3. 62·75 lb. Tan^{-1} 0·2 to plane.
4. ω = 68·75 lb. μ = 0·178, 47·5 lb., 22 lb.
5. (a) 115·6 lb., (b) 1 in 2·48.
6. 2·7 tons.
7. Force in rod = $1/\sqrt{3}$ tons, Thread torque = 0·0486 ton. in.
Thrust collar torque = 0·065 ton in. **Force = 51 lb. at 5 in. radius.**
8. 2275 ft.lb./min.
9. Horizontal force required = 1 ton.
α = 4° 33′, ϕ = 5° 43′. Screw torque = 0·181 ton in.
Total torque = 860 lb. in.
Work input/rev. = 5410 in. lb. Work output = 1120 in. lb.
Efficiency = 20·7%.
10. 200 lb.
11. μ = 0·0105, HP = 2·57.
12. (a) 927 lb. ft. ; (b) 1854 lb. ; (c) 0·069.

EXERCISES III : Page 81

1. V_A = 33·7 ; V_C = 72·2 ; 0·416.
2. 5 ft./sec.^{2}
3. 56·5 ft./sec. ; 5·57 miles ; 1·88 ft./sec.^{2} ; 105·6 ft./sec.
4. 0·733 ft./sec.^{2} ; 1·466 ft./sec.^{2} ; 440 ft.
5. 41·2 sec.
6. 2100 ft./sec. ; tan^{-1} $\frac{1}{5}$.
7. 77000 ft. ; 2065 ft./sec. at tan^{-1} 1·4 to horizontal.
8. (a) 6 ft./sec. to right ; 14 ft./sec. to left ; (b) 20 ft. to left ; 29 ft.
9. (a) 1 sec. and 3 sec. ; (b) 12 ft./sec.^{2} to left.
10. 10·3 rad./sec.^{2} ; 3240 ft./sec.^{2}
11. 70 rev./min. ; 1·13 min.
12. (a) 48·2 rad./sec.^{2} ; (b) 11·73 sec., 274 revs.
13. (a) 0·582 rad./sec.^{2} ; (b) 133 sec. ; (c) 297 revs. ; (d) 0·873 ft./sec.^{2}
14. 48 ft./sec. ; 64 ft.
15. (a) B relative to A, 8 m.p.h. 27·5° N. of W.
(b) 4·45 miles.
16. (a) 15 m.p.h. 29° N. of E. ; (b) $\sqrt{2}$ miles ; (c) $6\sqrt{2}$ minutes.
17. A = 206 miles, B = 219 miles.
18. 6·3 ft./sec. ; 10 ft./sec.

19. (a) 49 ft./sec. ; (b) 19·8 rad./sec. ; (c) 56·5 ft./sec. clockwise 15° to direction of piston velocity.

20. ∠BAD = 180° : (a) 1·14 ft./sec. ; (b) 3·04 rad./sec. ; 3·04 rad./sec. ∠BAD = 0° : (a) 2·2 ft./sec. ; (b) 5·88 rad./sec. ; 5·82 rad./sec.

21. 15°.

EXERCISES IV : Page 134

1. 0·228 lb.

2. 1 in 110 ; 806 h.p.

3. (a) 1·89 ft./sec.² ; (b) 5868 ft.lb.

4. 103·9 ft.

5. 4·25 tons ; 816 h.p.

6. 35·8 ft./sec. ; Max. velocity at height of 60 ft. = 38 ft./sec.

7. 2·025 lb. ft. ; average h.p. (based on 19000 rev./min.) = 7·34 ; 3600 lb.

8. 5·67 rad./sec. ; (Radial velocity relative to table = 1·89 ft./sec.). 29·7° with tangent.

9. (a) L. H. support : 90·2 lb. and 37·8 lb.
   R. H. support : 45·1 lb. and 18·9 lb.
   (b) 375 rev./min.

10. 2·205 ft. ; 116 ft./sec.²

11. 21·2 m.p.h.

12. (a) outer wheels, 6·533 tons ; inner wheels, 3·467 tons.
   (b) 81 m.p.h.

13. (a) 6·9° ; (b) 0·12.

14. 53 m.p.h. ; 3/8.

15. 40 ft./sec. ; 85·2 ft./sec.

16. 2·375 in. ; 99·9 tons.

17. 4·936 ton ft.

18. 2·84 ft./sec.

19. (a) 0·179 ft./sec.² ; (b) 8·18 sec. ; (c) 9·945 lb.

20. 2·14 Slug ft.² ; 6·84 in.

21. (a) 3 in. ; (b) 0·688 lb. in.

22. 393 sec. ; 4080.

23. (a) 672 ft.lb. ; (b) 420 rev./min. ; (c) 1·675 rad./sec.², 38·4 revs.

24. (a) 87·2 Slug ft.² ; (b) 172,400 ft.lb.

25. (a) 277 rev./min. ; (b) 30·3 lb. ft.

26. (a) 8/15 rad./sec.² ; (b) 6·01 in.

27. 58 rev./min.

28. 5·63 ft./sec.² ; 3763 lb.

29. (a) 17·65 rev./min. ; (b) 294·6 rev.

30. (a) 7·54 ft./sec. ; (b) 0·044 sec. ; (c) 16,070 + 3000 = 19,070 lb.

31. (a) 7·25 tons ; (b) $4\frac{11}{16}$ ft.ton.

32. (a) 4 ft./sec. ; (b) 13·5 ft.ton. ; (c) 40 ft.

33. 264 lb.

34. From vector diagram, relative velocity vector makes 135° with blade velocity at outlet (i.e. jet deflected 45° by stationary vane), relative velocity at inlet = relative velocity at outlet = 40 ft./sec. From isosceles triangle at outlet formed by relative velocity and blade velocity vectors jet is deflected through $22\frac{1}{2}°$ by moving vane. Mass of water *hitting* vane per sec. = 1·94 Slugs. Change in velocity in direction of motion = (80 − 68·28) ft./sec. Force equals **22·7 lb.** Change in velocity at right angles to motion = 28·28 ft./sec. Force at right angles = **54·86 lb.**

35. 103 rev./min. ; 291·5 ft.lb.

EXERCISES V : Page 158

1. 0·392 sec. ; 1·34 ft./sec.
2. 0·286 sec. ; 0·916 ft./sec.
3. 1·8 osc./sec. ; 1·37 ft./sec.
4. (a) 93·6 osc./min. ; (b) 2·31 ft./sec.
5. 0·159 sec. ; 3·28 ft./sec.
6. 3·69 in.
7. (a) 10 sec. ; 13·6 in. ; (b) 5·38 in./sec.^{2}
8. 169° 51′ ; 260 lb.
9. 18·4 lb.
10. 17,600 lb. ; 15,200 lb. ; 19·65 ft./sec.
11. (a) 34·2 lb. ; 5·85 ft./sec. ; (b) 51·3 lb. ; (c) 7·87 ft./sec.
12. (a) 1·43 ft. ; (b) 0·882 ft. ; (c) 0·1 sec.
13. (a) 2·53 ft. ; (b) 4·86 sec. ; (c) 2·75 lb.
14. 39·1 in.
16. (a) 2·215 sec. ; (b) 1·268 ft. from end ; (c) 2·05 sec.
17. 80·2 lb. ft.^{2} or 2·49 Slug ft.^{2}
18. Frequency $n = \dfrac{1}{2\pi}\sqrt{\dfrac{ga}{K_0{}^2}}$, For uniform plate Polar $I_G = \dfrac{d^4}{12} + \dfrac{d^4}{12} = \frac{1}{6}$ ft.^{4} where $d = 1$ ft.

By Parallel axis theorem $I_O = I_G + 1 \times \left(\dfrac{\sqrt{2}}{2}\right)^2 = \frac{2}{3}$ ft.^{4} where $a = \sqrt{2}/2$,

∴ $K_o{}^2 = \frac{2}{3}$ ft.^{2}. Hence $n = \mathbf{0·93}$ **osc./sec.**
19. 13·25 rads. ; 95·6 lb. ft.
20. (a) 1·71 Slug ft.^{2} ; (b) 17·5 in.

EXERCISES VI : Page 198

1. (a) 0·0278 in. ; (b) 5·6.
2. 12,160 tons/in.^{2}.
3. Steel : 20,000 lb./in.^{2} ; Copper : 12,000 lb./in.^{2}, 0·008 in./ft., 1·23 in. lb./ft. length.
4. 7·78 tons.
5. (a) $f_c = 7470$ lb./in.^{2}, $f_s = 19,500$ lb./in.^{2} ; (b) 0·0078 in. ; (c) 43·7 in. lb.

6. Cast iron = 1·85 tons/in.², Concrete = 0·461 tons/in.²
7. 0·0007 in.
8. Stress at end of steel rod = 4·08 tons/in.²
   Within sleeve the strains are the same for both materials.
   Ratio of stresses within sleeve to end = 1 : 1·9.
   Stress in brass = 0·86 tons/in.²
   Extension on 10″ steel = 0·00305 in.   Total extension = 0·00627 in.
9. (a) 14×10^{-4} ;   (b) 42,000 lb./in.²
10. (a) 50° F. ;   (b) 64,000 lb./in.²
11. Copper :   1330 lb./in.² ;   Steel :   1710 lb./in.² $(15 + 4455 \times 10^{-6})$
    inches.
12. Gunmetal : 8900 lb./in.² ;   Steel : 5000 lb./in.² ;   25,470 lb./in.²
13. 8000 lb. ;   9/16 in.
14. 109 lb./in.²
15. 1196 in.
16. Hoop Stress = 5000 lb./in.²   F. of S. for wall = 13·45.
    Force due to pressure on flanges = 62,840 lb.
    Stress in each bolt   = 4330 lb./in.²
    Comparison of factors of safety   = 1·21 : 1
17. 9-in. wide plate with 9 rivets.
18. 0·29 in.ton.
19. Steel, 3·35 tons/in.² ;   Concrete, 0·223 tons/in.² ;   0·897 in.ton.
20. Forces in tons :   DC = 9·03,   AC = 7·54,   AD = 3·96,   BD = 8·45 ;
    0·358 in.ton.
21. (a) Steel = 4·93 tons/in.²; C. I. = 2·65 tons/in.²
    Extension = 0·0265 in.
    (b) Steel = 0·098 in.ton, C. I. = 0·14 in.ton.
    (c) 7·395 tons.
23. (a) 70·8 in.lb. ;   (b) 3930 lb. ;   (c) 7860 lb. ;   0·018 in.
24. 0·224 in. ;   20,400 lb./in.²
25. 2·9 in.

EXERCISES VII : Page 229

1. R_A = 7·08 tons ; R_B = 9·92 tons.
2. $R_A = 22\frac{1}{15}$ tons ; $R_B = 16\frac{14}{15}$ tons.
3. R_A = 7·3   tons ; R_D = 9·7   tons ; M_A = 0 ; M_B = 73   ton   ft. ;
   M_C = 86·8 ton ft. ; M_D = 0.
4. R_A = 12 tons ; R_B = 12 tons.   Max. B.M. = 36 ton ft.
5. $R_A = 3\frac{1}{2}$ tons ; R_B = 12 tons.   Max. B.M. = 14 ton ft.
6. R_A = 19,500   lb. ; R_B = 13,500   lb. ; M_B = 32,000   lb.   ft. ; M_A
   = 50,000 lb. ft.   M. at conc. load = 20,000 lb. ft.
   Min. B.M. = 16,700 lb. ft. at 13·5 ft. from R.H.E.
7. $5\frac{1}{3}$ ft. from free end.
10. R_A = 7 tons ; R_E = 10 tons.   S.F. is zero at 7 ft. from A.   B.M. at
    this point = 24·5 ton ft.
11. R_A = 16 tons ; R_B = 9 tons.   Max. B.M. = 64 ton ft. at 8 ft. from A.

12. $R_A = 9.5$ tons ; $R_B = 40.5$ tons. Max. B.M. $= 90$ ton ft. at 19 ft. from A.

13. $R_A = 2\frac{1}{2}$ tons ; $R_C = 6\frac{1}{2}$ tons ; $M_D = 0$; $M_C = 10$ ton ft. ; $M_B = -8$ ton ft. ; $M_A = 0$. Max. B.M. $= -12\frac{1}{2}$ ton ft. at 10 ft. from A.

14. 12 cwt. 11 ft. from R. Max. B.M. $= 115.3$ cwt. ft. at 11 ft. from R.

15. $R_C = 32$ tons ; $R_B = 14$ tons 8 ft. from A.

16. 5·53 ft. and 14·47 ft. from A.

17. $R_A = 13\frac{1}{2}$ tons. $R_C = 14$ tons 21·32 ft. from A. Max. B.M. $= 64$ ton ft. at 10 ft. from A.

18. For the greatest bending moment to be a minimum the bending moment at the supports will be *numerically* equal to the bending moment at the centre span.
Supports placed 4·15 ft. from each end.
Points of contraflexure 5·875 ft. from ends.

19. Max. B.M. at centre $= 10$ ton ft.
Points of contraflexure 5·525 ft. from ends.
4·15 ft. from each end.

20. 1·5 tons/ft., 3·6 ft. from each end.

EXERCISES VIII : Page 252

1. 217 in. ; 0·01 ton.
2. 83·4 lb. in. ; 25,000 lb./in.²
3. 3·74 tons/in.² ; 1045 ft.
4. 7·85 ton in.
5. $d = 0.456$ in. ; $b = 0.912$ in. ; 30·17 lb. ft.
6. (a) 13 tons ; (b) 0·128 in.
7. $\frac{3}{8}$ in.
8. $b = 4.15$ in. ; $d = 12.45$ in.
9. (a) 1·43 tons ; Hollow/solid, 2·36 : 1.
10. 4·54 tons.
11. Max. B.M. $= 27.5$ ton ft. at 6-ton load. Max. Stress $= 33.72$ tons/in.² $R = 830$ ft.
12. 1·29 tons ; 3·56 tons/in.²
13. 220 lb./ft. ; 9175 lb./in.²
14. $5\frac{5}{8}$ ton ft. ; 2·5 tons/in.²
15. 35·1 lb./ft.
16. 5770 lb./in.² ; 5050 lb./in.²
17. 1 in. from face of flange, 3·79 in.⁴ ; 0·0947 ton/ft.
18. $f_c = 1380$ lb./in.² ; $f_t = 882$ lb./in.²
19. Max. B.M. $= 3900$ lb. ft. at 7·9 ft. from end farthest from concentrated load. $b = 3.4$ in. ; $d = 10.2$ in.
20. 1000 lb./in.² at centre ; 292 lb./in.² at 6·5 ft. from each end support.
21. Flanges, 7·1 tons/in.² ; Plates, 7·74 tons/in.² ; 50%.
22. Change in position of N.A. $= 0.42''$ towards 8″ flange, 47 in.⁴ Max. B.M. $= 60$ ton ft. Max. Stress $= 9.62$ tons/in.²
23. 1800 lb.

EXERCISES IX : Page 276

1. $5 \cdot 72 \times 10^6$ lb./in.²
2. $12 \cdot 13 \times 10^6$ lb./in.²
3. 55,500 lb. in. ; 10,500 lb./in.² ; 132.
4. $11 \cdot 2 \times 10^6$ lb./in.² ; 21,000 lb./in.² ; 1·97 in.
5. 1·452 tons/in.² ; 64·8.
6. 6530 lb./in.² ; 6·35 in.
7. 9020.
8. (i) 472 lb. in. ; (ii) 171·5 lb./in.² ; (iii) 0·0197°.
9. 1693 lb./in.²
10. 2170 ; 2·56°.
11. 8 in.
12. 39,720 lb./in.² ; 0·95°.
13. 21·7 in. ; 0·355°.
14. 4·05 tons/in.²
15. (a) 9·75 in. ; (b) 1·685°.
16. 2·4 : 1.
17. 2315 lb./in.² ; 3470 lb./in.²
18. 4·15 in. ; 4·47 in.
19. 2·5 in. ; 2·0 in. ; 2·41°.
20. 0·62 in. ; 8375 lb./in.²
21. $\frac{7}{8}$ in. ; 1875 lb./in.²
22. 9·9 in. ; $1\frac{3}{4}$ in.
23. 10·49 ; 0.000693 rads.
24. O.D. = 3·90 in. ; I.D. 2·34 in. ; 16,400 lb./in.² ; 8500 lb./in.²
25. Total torque transmitted by AB = 3 × Torque for BC = 95,400 lb. in. Max. Stress in AB = **7600 lb./in.²** Total h.p. = **908.** Total twist = 2·02°.
26. Solid = 4 in. ; O.D. of hollow = $4\frac{3}{4}$ in.
27. 2·2 in.
28. (a) 73·2 lb./in. ; (b) 980 lb.
29. $\frac{1}{2}$ in. ; 33·6 ft.
30. 19·5 ; 0·13 in.

EXERCISES X : Page 301

1. (a) 0·47 lb./in.² ; 1·33 lb./in.²
2. 2·75 in.
3. (a) 2725 lb. ; (b) 1900 lb. ; (c) 960 lb.
4. 90,400 lb. at 6·21 ft. from base.
5. 1061 lb. at 2·78 ft. from base.
6. 588 lb. at 1 in. below centre of plate.
7. 1764 lb. at 9/64 ft. below centre of plate.
8. 5770 lb.
9. 2666 lb. ; 2950 lb.
10. 3·78 ft.

11. 624 lb./ft.² ; 49 lb. ft.

12. $F = \omega A \bar{x} = 1275$ lb. With 0 in free surface OG = 6·06 ft.

I_G for triangular plate $= \dfrac{bh^3}{36}$. y_c along side = OG + 0·062 ft.

F acts at 0·81 ft. from A.

Force required at apex = **459 lb.**

13. 495 lb.

EXERCISES XI : Page 329

1. (a) 120·6 ft. lb./lb. ; 43·1 lb./in.²

2. 56 lb./in.²

3. 6·5 lb./in.²

4. $C_v = \sqrt{\dfrac{x^2}{4yH}}$ where $x = 33·5$ in. $y = 6·1$ in. $C_v = 0·98$.

$C_d = 0·637$. $C_c = 0·65$.

5. 215 g.p.m.

6. 6·36 in., pressure changes from 5 lb./in.² vac. to 3·93 lb./in.² vac., i.e. changes by 1·07 lb./in.²

7. 7020 gall./hr.

8. $C_d = 0·618$; $C_c = 0·64$; $C_v = 0·965$.

9. $C_d = 0·631$; $C_c = 0·679$; $C_v = 0·93$.

10. 615 gall./min.

11. 2·78 ft./sec. ; 4·14 lb./in.²

12. 27·8 lb./in.² ; 25·45 lb.

13. 12 in. ; 3400 gall./min.

INDEX

345